Daniel Hernández-Hernández • Florencia Leonardi •
Ramsés H. Mena • Juan Carlos Pardo Millán
Editors

Advances in Probability and Mathematical Statistics

CLAPEM 2019, Mérida, Mexico

Birkhäuser

Editors
Daniel Hernández-Hernández
Department of Probability and Statistics
Centro de Investigación en Matemáticas
Guanajuato, Mexico

Ramsés H. Mena
Departamento de Probabilidad
IIMAS-UNAM
Mexico City, Mexico

Florencia Leonardi
Instituto de Matemática e Estatística
Universidade de São Paulo
São Paulo, Brazil

Juan Carlos Pardo Millán
Department of Probability and Statistics
Centro de Investigación en Matemáticas
Guanajuato, Mexico

ISSN 1050-6977 ISSN 2297-0428 (electronic)
Progress in Probability
ISBN 978-3-030-85324-2 ISBN 978-3-030-85325-9 (eBook)
https://doi.org/10.1007/978-3-030-85325-9

Mathematics Subject Classification: 93E20, 90C40, 60J65, 60G51, 60F15, 60F10; Primary: 62C10, 60G57, 62F15, 99Z99; Secondary: 00A00, 62B10

This book is published under the imprint Birkhäuser, www.birkhauser-science.com, by the registered company Springer Nature Switzerland AG.
The registered company address is: Gewerbestrasse 11, 6330 Cham, Switzerland

Preface

The present volume contains contributions of the XV Latin American Congress of Probability and Mathematical Statistics (CLAPEM, by its acronym in Spanish), held at Merida, Mexico during December 2–6, 2019.

Endorsed by the Bernoulli Society, this event is the official meeting of the *Sociedad Latinoamericana de Probabilidad y Estadística Matemática* (SLAPEM) and it is the major event in probability and statistics in the region. It gathers an important number of researchers and students, predominantly from Latin America, serving as an ideal forum to discuss and to disseminate recent advances in the field, as well as to reveal the future of our profession.

Over nearly 40 years, the CLAPEMs have greatly contributed to the development of probability and statistics by promoting collaborations in the region as well as with the rest of the world. Previous editions were held in Caracas (1980, 1985, 2009), Montevideo (1988), Ciudad de México (1990), San Pablo (1993), Viña del Mar (1995, 2012), Córdoba (1998), La Habana (2001), Punta del Este (2004), Lima (2007), Cartagena (2014), and San José (2016).

On this occasion, the congress gathered scholars from over 20 countries and included a wide set of topics on probability and statistics. The scientific program included four plenary talks delivered by Gerard Ben Arous, Sourav Chatterjee, Thomas Mountdford, and Judith Rousseau. The event also benefited from eight semi-plenary talks given by Pablo Ferrari, Michele Guindani, Chris Holmes, Jean Michel Marin, Lea Popovic, and Fernando Quintana. The program also included two courses: "Hierarchical Bayesian Modeling and Analysis for Spatial BIG Data" by Sudipto Banerjee and "Sharpness of the phase transition in percolation" by Vincent Tassion, 10 thematic sessions, 21 contributed sessions, and several contributed talks and poster presentations.

The volume begins with the chapter by Andrade, Calvillo, Manrique, and Treviño where the authors present a probabilistic analysis of random interval graphs associated with randomly generated instances of the data delivery on a line problem (or DDLP). Angel and Spinka consider the infinite random geometric graph on a circle of circumference L, which is a random graph whose vertex set is given by a dense countable set in such circle, and find a dependency behavior on the rationality of L. The asymptotic behavior of four binary classification methods, when the dimension of the data increases and the sample sizes of the classes are fixed, are studied by Bolivar-Cime. A class of transport distances based on the Wassertein distances for random vectors of measures is considered by Catalano, Lijoi, and Prünster. The latter leads to a new measure of dependence for completely random vectors, and the quantification of the impact of hyperparameters in notable models for exchangeable time-to-event data. Gil-Leyva studies the construction of random discrete distributions, taking values in the infinite dimensional simplex, by means of latent random subsets of the natural numbers, which are then applied to construct Bayesian non-parametric priors. The connection between generalized entropies based on a certain family of α-divergences and the class of some predictive distributions is studied by Gutiérrez-Peña and Mendoza. A class of discrete-time stochastic controlled systems composed by a large population of N interacting individuals is considered by Higuera-Chan. The problem is studied by means of the so-called mean field model. Kouarfate, Kouritzin, and Mackay provide an explicit weak solution for the $3/2$ stochastic volatility model which is used to develop a simulation algorithm for option pricing purposes. Finally, León and Rouault revisit Wschebor's theorems on the a.s. convergence of small increments for processes with scaling and stationarity properties and then apply such results to deduce that large deviation principles are satisfied by occupation measures.

In summary, the high quality and variety of these chapters illustrate the rich academic program at the XV CLAPEM. It is worth noting that all papers were subject to a strict refereeing process with high international standards. We are very grateful to the referees, many leading experts in their own fields, for their careful and useful reports. Their comments were addressed by the authors, allowing to improve the material in this volume.

We would also like to extend our gratitude to all the authors whose original contributions appear published here as well as to all the speakers of the XV CLAPEM for their stimulating talks and support. Their valuable contributions encourage the interest and activity in the area of probability and statistics in Latin America.

We hold in high regard the editors of the series Progress in Probability: Davar Khoshnevisan, Andreas E. Kyprianou, and Sidney I. Resnick for giving us the opportunity to publish the symposium volume in this prestigious series.

Special thanks go to the Universidad Autónoma de Yucatán and its staff for its great hospitality and for providing excellent conference facilities. We are indebted to Rosy Dávalos, whose outstanding organizational work permitted us to concentrate mainly in the academic aspects of the conference.

The XV CLAPEM as well as the publication of this volume would not have been possible without the generous support of our sponsors and the organizing institutions: Bernoulli Society; Centro de Investigación en Matemáticas; Consejo Nacional de Ciencia y Tecnología; Facultad de Ciencias–UNAM; Gobierno de Yucatán; Google; Instituto de Investigaciones en Matemáticas Aplicadas y Sistemas–UNAM; Instituto de Matemáticas–UNAM; Universidad Autónoma de Chapingo; Universi-

Asymptotic Connectedness of Random Interval Graphs in a One Dimensional Data Delivery Problem

Caleb Erubiel Andrade Sernas, Gilberto Calvillo Vives, Paulo Cesar Manrique Mirón, and Erick Treviño Aguilar

Abstract In this work we present a probabilistic analysis of random interval graphs associated with randomly generated instances of the Data Delivery on a Line Problem (DDLP) (Chalopin et al., Data delivery by energy-constrained mobile agents on a line. In Automata, languages, and programming, pp. 423–434. Springer, Berlin, 2014). Random Interval Graphs have been previously studied by Scheinermann (Discrete Math 82:287–302, 1990). However, his model and ours provide different ways to generate the graphs. Our model is defined by how the agents in the DDLP may move, thus its importance goes beyond the intrinsic interest of random graphs and has to do with the complexity of a combinatorial optimization problem which has been proven to be NP-complete (Chalopin et al., Data delivery by energy-constrained mobile agents on a line. In Automata, languages, and programming, pp. 423–434. Springer, Berlin, 2014). We study the relationship between solvability of a random instance of the DDLP with respect to its associated interval graph connectedness. This relationship is important because through probabilistic analysis we prove that despite the NP-completeness of DDLP, there are classes of instances that can be solved polynomially.

Keywords Connectedness analysis · Data delivery problem · Mobile agents · Random interval graph

1 Introduction

The research presented in this work is in the intersection of several disciplines, Probability Theory, Computer Science, Operations Research, and Graph Theory. So, in this introduction we define the problem, and provide the basic concepts of computational complexity and random graphs that are needed to make the work

C. E. Andrade Sernas · G. Calvillo Vives · P. C. Manrique Mirón · E. Treviño Aguilar (✉)
Unidad Cuernavaca del Instituto de Matemáticas, Universidad Nacional Autónoma de México, Mexico City, Mexico
e-mail: erick.trevino@im.unam.mx

D. Hernández-Hernández et al. (eds.), *Advances in Probability and Mathematical Statistics*, Progress in Probability 79, https://doi.org/10.1007/978-3-030-85325-9_1

self-contained. We also include several references for those that may want to go deeper in the study of these subjects.

1.1 The Data Delivery on a Line Problem

The production of inexpensive, simple-built, mobile robots has become a reality nowadays, to the point that a swarm of mobile agents can be used for different tasks. A practical application would be, for instance, to use a swarm of drones to explore a cave and produce a map of it by collecting and sharing geospatial data whenever two drones meet. Or the use of a network of drones to deliver packages to customers from retail stores or courier services. The Data Delivery Problem is a mathematical abstraction of such scenarios and has been studied a lot lately, see e.g., [1, 3, 4, 8, 9, 12]. Here we deal with a specific Data Delivery Linear Problem (DDLP), namely where agents (robots) are constrained to move in a line. In this version, a set of n energy-constrained mobile agents are placed on positions $0 \leq x_i \leq 1, i = 1, \ldots, n$ on the unit interval and have ranges $\rho_i > 0$ (denoting the maximum length of a walk for each agent), the question is whether there is an order in which the mobile agents should be deployed to pick up the data at a point s called the source, and collectively move it to a predetermined destination called the target $t > s$. The first agent, in the order found, moves to the source, picks up the data and move it to the right according to its capacity. The second agent moves to the point where the first agent is, takes the data and move further to the right where a third agent comes to take over and so on until the data arrives, if possible, at its destination t. Observe that there are two cases for the movement of an agent. If the agent is to the left of the position d where the data is, it moves always to the right. First to pick up the data and then to move it further to the right. If it is to the right of d, then it has to move first to the left to reach d and then to the right as far as it can. In both cases, the agent covers the range $[x_i - a\rho_i, x_i + (1 - 2a)\rho_i]$ where $a = 0$ for the first case and $a = (x_i - d)/\rho_i$ for the second. This observation is key to define graphs associated to the problem. Observe also that the DDLP is a decision problem; that is to say, the answer is yes or no. In this framework now we can talk about the computational complexity of the problem.

This problem was introduced by Chalopin et al. [9], and it was shown to be NP-complete, although for instances where all input values are integers they gave a quasi-pseudo-polynomial time algorithm. In Sect. 2 we show how this problem is equivalent to a graph theoretical problem which in turn is analyzed using random graphs in Sect. 3. So let us first briefly recall some computational complexity and graph theoretical concepts.

1.2 Computational Complexity

This field has several roots: The seminal ideas of Jack Edmonds [13] about good algorithms and good characterizations; the foundational work of Stephen Cook [11] and Dick Karp [17], to name a few. The field is now represented by one of the millennium problems: *Is P=NP?*. A systematic treatment of the subject can be found in [15], [2].

A decision problem is a collection of instances (propositions) each of which is true or false. The DDLP is one such decision problems. Each instance is of the form: The set of agents $Q = \{(x_i, \rho_i), i = 1, \ldots, n\}$ can move the data from s to t. This proposition is true or false. The problem is to decide for each instance which is the correct answer. A decision problem is said to belong to the *class P* if there is an algorithm (Turing Machine) that decides correctly which is the answer and runs in polynomial time, which means that the number of steps the corresponding Turing Machine has to perform is bounded by a polynomial in the size of the instance (usually measured in bits). For the DDLP, the size of an instance is the number of bits required to store Q, s and t. A famous Decision Problem in the class P is to decide if a given set of linear inequalities has a solution. A decision problem belongs to the *class NP* if for every instance with answer YES, there is a polynomial algorithm to verify that the answer is correct. The DDLP is NP since for an affirmative answer it suffices to show the sequence in which the agents are used and check that effectively they move the data from s to t. This, of course can be done very efficiently. A decision problem S belongs to the class *NP-complete* if it is NP and any other NP problem can be reduced polynomially to S. Cook provided the first NP-complete problem (satisfiability) and then Karp [17] added a bunch of combinatorial problems to that class. Chalopin showed that DDLP is a NP-complete problem. While question *is P=NP?* is open, we do not know if the NP complete problems can be solved by polynomial time algorithms. At present, the generalized belief is that $P \neq NP$. If this is so, NP problems will remain as difficult problems, including the DDLP. One of the drawbacks of this theory is that it does not recognize clearly that a large amount of an NP-complete problem can be solved efficiently. The recognition of that fact has motivated the use of probability to asses fringes of "easy" instances and thus isolate the really hard instances of a problem. A good example of this methodology is in [6]. Our work follows that path. In Sect. 5 we interpret the results presented in Sect. 3 in this sense.

1.3 Graphs

For the purpose of this work it is sufficient to deal with simple graphs and so we omit the adjective. A graph consists of a set V whose elements are called vertices and a collection of subsets of V of cardinality 2 which we denote by E and call them edges. If a vertex v belongs to an edge e we say that v is an extreme of e. Clearly every edge has two extremes. If u and v are the two extremes of an edge we say that they are adjacent. A simple path (or path) is a sequence $v_0, e_1, v_1, e_2, v_2, \ldots, v_{n+1}, e_n, v_n$ in which all vertices $v_i, i = 1, \ldots, n$ are different and v_i and v_{i+1} are the extremes of edge e_i. Such path is said to connect vertices v_0 and v_n; when $v_0 = v_k$, the graph is a cycle, denoted by C_k. A graph is connected if for every pair of vertices u and v of the graph, there exists a path connecting them. Given a set V of real closed intervals in the real line we can construct a graph in the following way: The set of vertices is V and two vertices are adjacent (form an edge) if they intersect as intervals. A graph constructed in this way is called an **interval graph**. The graphs that can be constructed in this way are a small part of all possible graphs. For example, any cycle C_k with $k > 3$ is not an interval graph. However they are nice graphs to work with because they have remarkable properties. There are several ways in which a collection of intervals can be defined. We are interested in two of them.

Definition 1 The symmetric model of a interval graph is generated by intervals I_i defined by its center x_i and its radius ρ_i in the form $I_i = [x_i - \rho_i, x_i + \rho_i]$. The interval graph associated to the symmetric model will be denoted by $G_S(Q)$ where $Q = \{(x_i, \rho_i), i = 1, \ldots n\}$.

The **asymmetric model** is obtained from instances of DDLP considering $I_i = [x_i - a_i\rho_i, x_i + (1 - 2a_i)\rho_i]$ defined in Sect. 1.1. It will be denoted by $G_A(Q, a)$, where again $Q = \{(x_i, \rho_i), i = 1, \ldots n\}$ and a is a vector in the unit box of $\mathbb{R}^n$.

A standard reference in graph theory is [7].

1.4 *Random Interval Graphs*

Random graphs were firstly introduced by Gilbert [16], but the work of Erdős and Rényi [14] set the basis for the study of their evolution (a comprehensive study on random graphs can be found in [6]). Later on, Cohen studied the asymptotic probability that a random graph is a unit interval graph [10], and Scheinermann replicated the ideas of Erdős and Rényi to study the evolution of random interval graphs [18].

To continue our discussion, we introduce a probabilistic framework, so let $(\Omega, \mathcal{F}, \mathbb{P})$ be a probability space with $\mathcal{F}$ a σ-algebra on the set of scenarios Ω and $\mathbb{P}$ a probability measure on $\mathcal{F}$. All random variables and our asymmetric model will be defined there.

Scheinerman [18] obtained random interval graphs, using the symmetric model, by defining two random variables for each interval, the centers $\{x_i\}_{i=1}^n$ and the radii of the intervals $\{\rho_i\}_{i=1}^n$, the centers with uniform distribution in [0,1] and the radii also uniform in the interval $[0, r]$ with $r < 1$, so intervals are constructed as $[x_i - \rho_i, x_i + \rho_i]$. We use the asymmetric model fixing the parameters $\{a_i\}_{i=1}^n$, $a_i \in [0, 1]$ and considering $2n$ independent random variables, the locations $\{x_i\}_{i=1}^n$ (all identically distributed uniformly in $[0, 1]$), and the ranges $\{\rho_i\}_{i=1}^n$ (all identically distributed uniformly in $[0, r]$). The intervals are $[x_i - a_i \rho_i, x_i + (1 - 2a_i)\rho_i]$, which are substantially different to the intervals of the symmetric model. First they are asymmetric, second they are shorter as specified by the parameter a_i which represents the percentage of a mobile agent's energy used to move backwards to pick up the load before going forward with it, thus it determines the behavior of the agent. The asymptotic analysis presented in this work is analogous to the one of Scheinermann's, but here we consider an asymmetric model which is more general. Moreover, it is related to a NP-complete combinatorial problem which makes the new model interesting from the point of view of computational complexity, a fact that we show in this work.

2 Graph Theoretical Formulation of the Data Delivery on a Line Problem

In this section we show how the DDLP is equivalent to an existence problem for interval graphs. The transformation from one another is completely general and does not assume that the data have been randomly generated. The equivalent graph theoretical decision problem is the following: Given a finite collection $Q = \{(x_i, \rho_i), i = 1, \ldots n\}$ of points in $\mathbb{R}^2$ decide if there exists a vector $a = (a_1, \ldots, a_n)$ such that the interval graph $G_A(Q, a)$ defined by the asymmetric model is connected. This problem will be called the *Existential Connectedness Problem* (ECP).

We say that two decision problems are equivalent if for every instance of one there is an instance of the other such that both have an affirmative answer or both have a negative answer.

Theorem 1 *DDLP and ECP are equivalent.*

Proof First we will show that for every instance (Q, s, t) of DDLP there exists an instance of ECP whose solution conforms with the one of the DDLP instance. First, discard all points of Q such that either $x_i + \rho_i < s$ or $x_i - \rho_i > t$ then add two new points $(s, 0)$ and $(t, 0)$. Call this new set of points Q' which defines an instance of the ECP. A positive answer to this instance comes with a vector a^* such that $G_A(Q', a^*)$ is a connected graph. So there exists a path T in $G_A(Q', a^*)$ from the vertex defined by $(s, 0)$ to the vertex defined by $(t, 0)$. The interior vertices of T correspond to points of Q and so to agents in the DDLP instance. The order in which the vertices of T are traversed from s to t define the order in which the agents have to be deployed. The adjacency of consecutive vertices of T guarantee that the corresponding agents can get in touch to pass the data from one to the next. So the existence of an affirmative solution of ECP translates into a solution of DDLP.

It remains to be proven that if the solution to the ECP instance is negative so is the answer to the DDLP. The contrapositive of this is that any YES answer to an instance of DDLP translates into an affirmative answer of ECP. To prove this let

(Q, s, t) be an instance of DDLP with a true answer and a sequence $r_{i_1}, \ldots, r_{i_k}$ of agents that carry the data from s to t. This behaviour of the agents in the sequence define a set of parameters $a_{i_1}, \ldots, a_{i_k}$. For the rest of the agents, which do not play any role in the transportation of the data, let $a_i = 0$ if $x_i < t$ and $a_i = 1$ if $x_i \geq t$. We claim that the graph $G_A(Q, a)$ is connected. The parameters $a_{i_1}, \ldots, a_{i_k}$ define a set of intervals, a path T' in $G_A(Q, a)$, that cover the segment $[s, t]$. We will show that every other interval intersects T'. For an agent r_i such that $x_i < t$ the corresponding interval is $[x_i, x_i + \rho_i]$, if $x_i \geq s$, then x_i is in $[s, t]$ and so it intersects some interval of T'; if $x_i < s$, then since irrelevant agents have been removed $x_i + \rho_i \geq s$ and therefore the interval intersects T' too. The case $x_i \geq$ t is resolved similarly. □

The next result is a direct consequence to Theorem 1. We omit the details of the proof.

Corollary 1 *ECP is NP-complete*

Theorem 1 shows that DDLP, which is NP-complete, is reducible to ECP. Moreover the reduction is polynomial since to transform a DDLP given by (Q, s, t) to a ECP given by Q' it is only needed to compute which robots (x_i, ρ_i) satisfy $x_i - \rho_i < s$ or $x_i + \rho_i > t$; and to include the points $(s, 0)$, $(t, 0)$. This can be done in linear time assuming arithmetic operations are performed in constant time. The equivalence between DDLP and ECP allow us to deal with the graph theoretical formulation and use the results and ideas of Random Graphs in order to obtain some asymptotic results in the ECP and therefore the DDLP problems. Specifically, parameters' domain can be partitioned into regions, one in which the problems can efficiently be solved. The relevance of the connectedness of an interval graph in relation to the solvability of its associated DDLP instance cannot be overstated. That is, if for any DDLP instance (Q, s, t) there exists a vector a in the unit box such that $G_A\left(Q', a\right)$ is connected, then the DDLP instance is solvable, the converse is certainly not true, as shown in the counterexamples of Fig. 1. On the other hand, given a DDLP instance (Q, s, t), if for all possible vectors a in the unit box none of the associated interval graphs $G_A\left(Q', a\right)$ is connected, it is then assured that the instance is not solvable. This last statement is summarized in the following corollary.

Corollary 2 *Given a DDLP instance (Q, s, t) if its associated symmetric interval graph $G_S\left(Q'\right)$ is disconnected, the instance is not solvable.*

Proof For every vector a in the unit box we have that

$$[x_i - a y_i, x_i + (1 - 2a)\, y_i] \subset [x_i - y_i, x_i + y_i]\,.$$

Now, recall that $Q' = Q \cup \{(s, 0)\,, (t, 0)\}$, so if $G_S\left(Q'\right)$ is disconnected, it means that for every vector a in the unit box $G_A\left(Q', a\right)$ is disconnected as well. This means that ECP is not solvable, so by Theorem 1 DDLP is not solvable. □

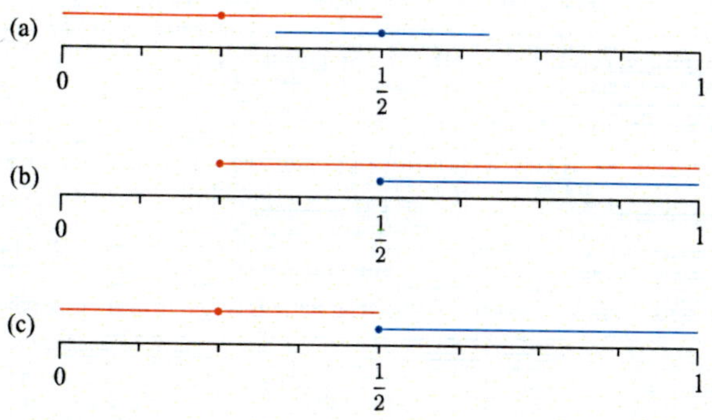

Fig. 1 For this example we take $Q = \left\{\left(\frac{1}{4}, \frac{3}{4}\right), \left(\frac{1}{2}, \frac{1}{2}\right)\right\}$. In panel (**a**) we have $a = \left(\frac{1}{3}, \frac{1}{3}\right)$, in (**b**) $a = (0, 0)$, and in (**c**) $a = \left(\frac{1}{3}, 0\right)$. In the first two cases $G_A\left(Q', a\right)$ is disconnected, but in the third case it is connected and therefore solvable

3 Probabilistic Analysis of $G_A(Q, a)$

We denote as usual the degree of a vertex v_i by $d(i)$. We write $v_i \not\sim v_j$ when there is no edge joining the vertices v_i and v_j. We denote by $v_i^+ = x_i + (1 - 2a)\rho_i$ (resp. $v_i^- = x_i - a\rho_i$) the right (resp. left) boundary of v_i. In this section we analyze the connectedness of random interval graphs of the form $G_A(Q, a)$. The approach that we follow is to fix a and analyze the asymptotic behavior of the connected components of $G_A(Q, a)$. In order to do so, we first introduce random variables that count the connected components of a random graph $G_A(Q, a)$. In this section and the sequel we will specialize to the case in which a is a scalar, that is $a_i = a$ for all i. Furthermore, we assume $a \in [0, 1/2]$. Estimations with this specification will already yield lower bounds for the probability of solvability due to Theorem 1 which is our main goal.

For each $i, i = 1, \ldots, n$ let

$$X_i := \begin{cases} 1 & \text{if the right end point of } v_i \text{ is contained in no other interval } v_j,\ j \neq i, \\ 0 & \text{otherwise.} \end{cases}$$

The family of random variables $X_1, \ldots, X_n$ indicates where a connected component ends. For example, assume that $X_1 = X_2 = 1$ and $X_i = 0$ for $i > 2$. Assume without loss of generality that the interval v_1 is on the left of the right end point of v_2. Since $X_1 = 1$ the right end point of v_1 is not included in any other interval, implying that all the intervals on the left of v_1^+ are disjoint from all other intervals on its right. Thus, there are at least two components. In order to see that there are exactly two components, assume that there is one more. For this component denote by v_j the interval such that v_j^+ attains the supremum over all right ends of intervals in the components. Then, necessarily $X_j = 1$ with $j > 2$. This is a contradiction. Then, there exists exactly two components.

We also define

$$X(n) := \sum_{i=1}^{n} X_i$$

is the random variable that counts the number of connected components of $G_A(Q, a)$. Note that $\mathbb{P}(X(n) = 0) = 0$. The distribution of $X(n)$ depends on the parameter r. However, we do not write this explicitly in order not to overload notation. In our first asymptotic result, we let the numbers of vertices n goes to infinite while at the same time shrinking the range of the intervals (the parameter r). The theorem gives the right trade-off for those processes in order to have connectedness with high probability.

Theorem 2 *Let* $\beta := \frac{2}{1-a}$ *and* $r(n) := \beta \frac{1}{n}(\log(n) + c)$. *Then,*

$$\liminf_{n\to\infty} \mathbb{P}(X(n) = 1) \geq 1 - e^{-c}. \tag{1}$$

Proof Denote by $v_i^+ = x_i + (1 - 2a)\rho_i$ the right point of the interval v_i. Let

$$B(i) := \{\omega \mid v_i^+(\omega) \in \cup_{k\neq i} v_k\}$$

be the event in which the right boundary of the interval v_i is included in at least one of the other intervals. The event $\{X(n) = 1\}$ satisfies

$$\{X(n) = 1\} = \bigcup_{k=1}^{n} \bigcap_{i\neq k} B(i).$$

Hence

$$\mathbb{P}(X(n) = 1) = \mathbb{P}\left(\bigcup_{k=1}^{n} \bigcap_{\substack{i=1\\ i\neq k}}^{n} B(i)\right) \geq \mathbb{P}\left(\bigcap_{i=2}^{n} B(i)\right).$$

Moreover

$$1 - \mathbb{P}\left(\bigcap_{j=2}^{n} B(j)\right) = \mathbb{P}\left(\left\{\bigcap_{j=2}^{n} B(j)\right\}^c\right) \leq \sum_{j=2}^{n} \mathbb{P}\left(B^c(j)\right) = (n-1)\mathbb{P}\left(B^c(2)\right).$$

Hence:

$$\mathbb{P}(X(n) = 1) \geq 1 - (n-1)\mathbb{P}\left(B^c(2)\right). \tag{2}$$

So now we estimate $\mathbb{P}(B^c(2))$. The right end point v_i^+ is a random variable and we denote by μ it probability distribution function under P. Note that the distribution does not depend on i. Then,

$$\mathbb{P}\left(B^c(2)\right) = \int_0^{1+(1-2a)r} (1 - \mathbb{P}(t \in v_2)))^{n-1} \mu(dt),$$

due to the independency of the intervals. For $a < \frac{1}{2}$ the random variable v_i^+ is the sum of two independent uniform distributions and then, we can easily see that μ is a distribution concentrated on $[0, 1 + r(1-2a)]$ with density

$$f(x) = \begin{cases} \frac{x}{r(1-2a)} & [0, r(1-2a)] \\ 1 & [r(1-2a), 1] \\ \frac{-x+1+r(1-2a)}{r(1-2a)} & [1, 1+r(1-2a)] \\ 0 & otherwise. \end{cases}$$

For $t \in [0, 1 + r(1-2a)]$ let $h(t, x, \rho)$ be the indicator function of $\{x - a\rho \leq t\} \cap \{x + (1-2a)\rho \geq t\}$ and $g(t, \rho) := \left\{(t + a\rho) \wedge 1 - (t - (1-2a)\rho)^+ \wedge 1\right\}$. We have

$$\begin{aligned} \mathbb{P}(t \in v_2) &= \frac{1}{r}\int_0^r \int_0^1 h(t, x, \rho)\mu(dx)d\rho \\ &= \frac{1}{r}\int_0^r g(t, \rho)d\rho. \end{aligned}$$

For $t \in [0, (1-2a)r]$ the function g simplifies to $g(t, \rho) = t + a\rho - (t - (1-2a)\rho)1_{\{t>(1-2a)\rho\}}$. Hence

$$\begin{aligned} \mathbb{P}(t \in v_2) &= t + \frac{1}{2}ar - \frac{1}{r}\left[t\rho - \frac{1}{2}(1-2a)\rho^2\right]_0^{\frac{t}{1-2a}} \\ &= t + \frac{1}{2}ar - \frac{1}{2r(1-2a)}t^2. \end{aligned}$$

Note that $\mathbb{P}(t \in v_2) \geq \frac{1}{2}ar$. Then

$$\begin{aligned}
&\frac{1}{(1-2a)r}\int_0^{(1-2a)r}(1-\mathbb{P}(t \in v_2)))^{n-1}\, tdt \\
\leq &\frac{1}{(1-2a)r}\int_0^{(1-2a)r}\left(1-\frac{1}{2}ar\right)^{n-1} tdt \\
= &\left(1-\frac{1}{2}ar\right)^{n-1}(1-2a)\,r.
\end{aligned}$$

For $x \in \mathbb{R}$ we will make use of the following basic inequality:

$$\left(1-\frac{x}{n}\right)^n \leq e^{-x}.$$

Then

$$\begin{aligned}
&(n-1)\left(1-\frac{1}{2}ar(n)\right)^{n-1}(1-2a)\,r(n) \\
\leq &n\exp\{-\frac{(n-1)a}{2}r(n)\}r(n) \\
= &\left(\frac{1}{n}\right)^{\frac{a\beta}{2}\frac{n-1}{n}}\exp\left\{-\frac{a\beta}{2}\frac{n-1}{n}c\right\}\beta(\log(n)+c).
\end{aligned}$$

Thus, the contribution of $(n-1)\int_0^{(1-2a)r(n)}(1-\mathbb{P}(t \in v_2)))^{n-1}\,\mu(dt)$ to $(n-1)\mathbb{P}(B^c(2))$ goes to zero asymptotically. In a similar fashion we can see that

$$\lim_{n\to\infty}(n-1)\int_{1-ar(n)}^{1+(1-2a)r(n)}(1-\mathbb{P}(t \in v_2)))^{n-1}\,\mu(dt) = 0.$$

For $t \in [(1-2a)r, 1-ar]$ we have

$$\begin{aligned}
g(t,\rho) &= (t+a\rho)\wedge 1-(t-(1-2a)\rho)^+\wedge 1 \\
&= t+a\rho-t+(1-2a)\rho \\
&= (1-a)\rho.
\end{aligned}$$

Hence, for $t \in [(1-2a)r, 1-ar]$

$$\begin{aligned}
\mathbb{P}(t \in v_2) &= \frac{1}{r}\int_0^r(1-a)\rho d\rho \\
&= \frac{1}{2}(1-a)r.
\end{aligned}$$

Moreover

$$\begin{aligned}(n-1)\int_{(1-2a)r}^{1-ar}(1-\mathbb{P}(t\in v_2))^{n-1}\,\mu(dt) =&(n-1)(1-\frac{1}{2}(1-a)r)^{n-1}(1-(1-a)r)\\ \leq&(n-1)\exp\left\{-\frac{1}{2}(1-a)nr(n)\right\}\\ \leq&e^{-c}.\end{aligned}$$

We have proved the theorem. □

Remark 1 Theorem 2 tells us that the probability that the interval graph is connected is bounded from below as $n \to \infty$ and $r(n)$ converges to zero but not too fast. The intuition behind this threshold comes from the order statistics of uniform random variables. If we generate n random points with uniform distribution in the interval $[0, 1]$, and call $U_{(k)}$ the k-th order statistic, then $\frac{U_{(k)}}{k/n} \to 1$ in probability as $k \to \infty$; see e.g., [5, Thm. 1.4]. That is to say, the random points tend to be homogeneously spattered as n grows, such that the spacing between consecutive points tends to be $1/n$. It might be reasonable to think that if r decreases as $\approx \frac{1}{n}$ the symmetric intervals $[x_i - \rho_i, x_i + \rho_i]$ would yield a connected component. Still, we need a little adjustment to account for the random lengths of each interval. In fact $\frac{\log n}{n}$ is the threshold to ensure connectivity in the symmetric model [18]. Our model is not symmetric but rather the intervals are of the form $[x_i - a\rho_i, x_i + (1-2a)\,\rho_i]$, and in Theorem 2 we showed that the connectivity threshold must be adjusted to account for this asymmetry.

The random variable $X(n)$ counts the number of connected components, so $X(n) = 1$ means that the interval graph is connected. If the interval graph associated to a DDLP instance is a single connected component, with the additional requirements that there exists intervals intersecting the boundaries, the instance is solvable due to Theorem 1. In the next theorem we show that the parametrization of r as function of n in Theorem 2 is optimal. For the statement of the theorem we introduce some random variables. Let $Y(n) := \sum_{i=1}^{n} Y_i$ where

$$Y_i := \begin{cases} 1 & \text{if } d(i) = 0 \text{ and } x_i \in \left[\frac{1}{3}, \frac{2}{3}\right] \\ 0 & \text{otherwise.} \end{cases}$$

Observe that $Y(n)$ counts the number of intervals $v_i = [x_i - a_i\rho_i, x_i + (1-2a_i)\,\rho_i]$ for which $x_i \in [1/3, 2/3]$ with isolated associated vertices in the interval graph.

Theorem 3 *Let β be as in Theorem 2. Let $\alpha(n) := \frac{1}{\beta}\frac{nr(n)}{\log n}$. If $\alpha := \lim_{n\to\infty}\alpha(n) < 1$ then*

$$\lim_{n\to\infty}\frac{1}{\log(n)}\mathbb{E}[Y(n)] = \infty \tag{3}$$

and

$$\lim_{n\to\infty} \mathbb{P}\left(Y(n) \geq \frac{1}{\log(n)}\mathbb{E}[Y(n)]\right) = 1. \tag{4}$$

In particular, with high probability all graph instances are not connected.

Proof We start with the idea of the proof for (4). Let $\sigma^2(n) := \boldsymbol{variance}(Y(n))$. For $\epsilon > 0$, Chebyshev's inequality $\mathbb{P}(|Y(n) - \mathbb{E}[Y(n)]| \geq \epsilon) \leq \frac{\sigma^2(n)}{\epsilon^2}$ yields $1 - \frac{\sigma^2(n)}{\epsilon^2} \leq \mathbb{P}(-\epsilon + \mathbb{E}[Y(n)] \leq Y(n))$. In particular, for $\epsilon(n) := (1 - \frac{1}{\log(n)})\mathbb{E}[Y(n)]$

$$\frac{\sigma^2(n)}{\epsilon^2(n)} = (1 - \frac{1}{\log(n)})^{-2}\left(\frac{\mathbb{E}[Y^2(n)]}{\mathbb{E}^2[Y(n)]} - 1\right),$$

and we will show

$$\lim_{n\to\infty} \frac{\mathbb{E}[Y^2(n)]}{\mathbb{E}^2[Y(n)]} = 1.$$

This establishes (4) since

$$\{-\epsilon(n) + \mathbb{E}[Y(n)] \leq Y(n)\} = \{Y(n) \geq \frac{1}{\log(n)}\mathbb{E}[Y(n)]\}.$$

Next we compute $\mathbb{E}(Y_i)$. We have

$$\begin{aligned}
\mathbb{E}(Y_i) &= \mathbb{P}(d(i) = 0 \text{ and } x_i \in [1/3, 2/3]) \\
&= \mathbb{P}(v_i \not\sim v_k \text{ for all } k \neq i \text{ and } x_i \in [1/3, 2/3]) \\
&= \int \mathbf{1}_{\{x_i \in [1/3, 2/3]\}} (\mathbb{P}(v_i \not\sim v))^{n-1}\, d\mu(v_i),
\end{aligned}$$

where μ is the distribution associated to the random vertex v_i. Note that

$$\mathbb{P}(v_i \not\sim v \mid \rho = t) = 1 - (1-a)\rho_i - (1-a)t.$$

Now we have

$$\begin{aligned}
&\int \mathbf{1}_{\{x_i \in [1/3, 2/3]\}} (\mathbb{P}(v_i \not\sim v))^{n-1}\, d\mu(v_i) \\
&= \int \mathbf{1}_{\{x_i \in [1/3, 2/3]\}} \left(1 - (1-a)\rho_i - (1-a)\frac{r}{2}\right)^{n-1} d\mu(v_i) \\
&= \left[\int_{1/3}^{2/3} dx_i\right]\left[\frac{1}{r}\int_0^r \left[1 - (1-a)\rho_i - (1-a)\frac{r}{2}\right]^{n-1} d\rho_i\right] \\
&= \frac{1}{3} \cdot \frac{\left[1 - \frac{1-a}{2}r\right]^n - \left[1 - \frac{3(1-a)}{2}r\right]^n}{n(1-a)r}.
\end{aligned} \tag{5}$$

From (5), we determine $\mathbb{E}[Y(n)]$ by

$$\mathbb{E}(Y(n)) = \frac{1}{3}\frac{\beta}{2}\frac{\left[1-\frac{1}{\beta}r(n)\right]^n - \left[1-3\frac{1}{\beta}r(n)\right]^n}{r}$$

$$= \frac{1}{6}\beta\frac{n}{\log(n)}\frac{\left[1-\frac{1}{\beta}\frac{nr(n)}{\log(n)}\frac{\log(n)}{n}\right]^n - \left[1-3\frac{1}{\beta}\frac{nr(n)}{\log(n)}\frac{\log(n)}{n}\right]^n}{\frac{nr(n)}{\log(n)}}$$

$$= \frac{1}{6}\frac{1}{\alpha(n)}\frac{n}{\log(n)}\left\{\left[1-\alpha(n)\frac{\log(n)}{n}\right]^n - \left[1-3\alpha(n)\frac{\log(n)}{n}\right]^n\right\}$$

$$\approx \frac{n^{1-\alpha}-n^{1-3\alpha}}{6\alpha\log(n)}. \tag{6}$$

Let us estimate $\mathbb{E}(Y^2(n))$. To this end, note first that $\mathbb{E}(Y^2(n)) = \mathbb{E}(Y(n)) + n(n-1)\mathbb{E}(Y_1Y_2)$ and $\mathbb{E}(Y_1Y_2) = \mathbb{P}(Y_1Y_2 = 1, v_1 \not\sim v_2)$. For this last probability, we have

$$\mathbb{P}(Y_1Y_2 = 1, v_1 \not\sim v_2) =$$
$$\int \mathbf{1}_{\{v_1\not\sim v_2\}}\mathbf{1}_{\{x_1,x_2\in[1/3,2/3]\}}\mathbb{P}(v_1, v_2 \not\sim v_k \text{ for all } k \neq 1,2)\, d\mu(v_1, v_2). \tag{7}$$

Note for k_0 fixed, $\mathbb{P}(v_1, v_2 \not\sim v_k \text{ for all } k \neq 1,2) = \left[\mathbb{P}\left(v_1, v_2 \not\sim v_{k_0}\right)\right]^{n-2}$. Conditioning

$$\mathbb{P}\left(\{v_1, v_2 \not\sim v\} \cap \left\{\left|v_1^{\pm} - v_2^{\mp}\right| \geq (1-a)\rho\right\} \mid \rho\right) = 1-(1-a)\rho_1-(1-a)\rho_2-2(1-a)\rho \tag{8}$$

$$\mathbb{P}\left(\{v_1, v_2 \not\sim v\} \cap \left\{\left|v_1^{\pm} - v_2^{\mp}\right| < (1-a)\rho\right\} \mid \rho\right) = 1-(1-a)\rho_1-(1-a)\rho_2-(1-a)\rho. \tag{9}$$

Note that $\left|v_1^{\pm} - v_2^{\mp}\right| < (1-a)\rho$ happens with probability $O(r)$ as $r \to 0$. Wrapping all together

$$\mathbb{P}(Y_1Y_2 = 1, v_1 \not\sim v_2) = (1-O(r))\mathbb{P}\left(Y_1Y_2 = 1, v_1 \not\sim v_2, \left|v_1^{\pm} - v_2^{\mp}\right| \geq (1-a)\rho\right)$$
$$= (1-O(r))\int \mathbf{1}_{\{x_1,x_2\in[1/3,2/3]\}}\left[\mathbb{P}\left(v_1, v_2 \not\sim v, \left|v_1^{\pm} - v_2^{\mp}\right| \geq (1-a)\rho\right)\right]^{n-2} d\mu(v_1, v_2)$$
$$= (1-O(r))\int \mathbf{1}_{\{x_1,x_2\in[1/3,2/3]\}}\left[1-(1-a)\rho_1-(1-a)\rho_2-2(1-a)\frac{r}{2}\right]^{n-2} d\mu(v_1, v_2)$$
$$= (1-O(r))\left[\int \mathbf{1}_{\{x_1,x_2\in[1/3,2/3]\}} d\mu(x_1, x_2)\right] \times \tag{10}$$

$$\left[\frac{1}{r^2}\int_0^r\int_0^r\left[1-(1-a)\rho_1-(1-a)\rho_2-2(1-a)\frac{r}{2}\right]^{n-2}d\rho_1 d\rho_2\right]$$
$$=(1-O(r))\frac{1}{9}\cdot\frac{[1-(1-a)r]^n-2\,[1-2(1-a)r]^n+[1-3(1-a)r]^n}{n(n-1)r^2(1-a)^2}. \tag{11}$$

Since $\alpha < 1$ and $\alpha\beta = \lim_{n\to\infty}\frac{nr}{\log n}$, from (5) and (10), we have

$$n(n-1)\mathbb{E}(Y_1Y_2) = (1-O(r))\frac{1}{9}\cdot\frac{[1-(1-a)r]^n-2\,[1-2(1-a)r]^n+[1-3(1-a)r]^n}{r^2(1-a)^2}$$
$$\approx \frac{n^2}{36\alpha^2\log^2(n)}\left[n^{-2\alpha}-2n^{-4\alpha}+n^{-6\alpha}\right]. \tag{12}$$

From (6) and (12), we get

$$\frac{\mathbb{E}(Y^2(n))}{(\mathbb{E}(Y(n)))^2}\to 1,$$

when $n\to\infty$. □

Now returning to our original motivation of a DDLP, Theorem 3 does not yet fully address the issue to determine if a connected interval graph is synonym of solvability of the DDLP. If the interval graph associated to a DDLP instance is a single connected component, but none of its intervals actually intersect the boundaries, we cannot say that the instance is solvable. However, the next result shows that the boundaries will be reached with high probability.

Theorem 4 *Let $r(n)$ be as in Theorem 2. Let $G(n)$ be the event that the associated interval graph is connected. Define the random variables*

$$R(n) = \max_{i=1\ldots,n}\{x_i\},\, L(n) = \min_{i=1\ldots,n}\{x_i\} \tag{13}$$
$$R'(n) = \max_{i=1\ldots,n}\{x_i + (1-2a)\,\rho_i\},\, L'(n) = \min_{i=1\ldots,n}\{x_i - a\rho_i\}. \tag{14}$$

Then $R(n)-L(n)$ has Beta distribution with parameters $(n-1, 2)$. Furthermore, for $\epsilon\in(0,1)$

$$\lim_{n\to\infty}\mathbb{P}\left(R'(n)-L'(n)\le\epsilon \mid G(n)\right) = 0. \tag{15}$$

Proof To simplify notation we write R, L and R', L'. Recalling $\{x_i\}_{i=1\ldots,n}$ are independent and identically distributed in [0, 1], it is straightforward to check that $\mathbb{P}(L\le t) = 1-(1-t)^n$ so that its density function is $f_L(t) = n(1-t)^{n-1}$. Analogously, $\mathbb{P}(R\le t) = t^n$ and its density function is $f_R(t) = nt^{n-1}$. Now, because L and R are not independent we cannot compute the density of $R-L$ as the convolution of R and $-L$, so we proceed as follows.

For $d\in(0,1)$ we want to compute $\mathbb{P}(R-L\le d)$. To this end, let do some conditioning on L. Let $t\in[0,1]$ denote the actual minimum of $\{x_i\}_{i=1\ldots,n}$. We must distinguish two cases that are mutually exclusive. The first case is $t < 1 - d$, so given that $L = t$, the probability that the remaining $n - 1$ points fall within the interval $[t, t + d]$ is given by $\left(\frac{d}{1-t}\right)^{n-1}$. The second case is $t \geq 1-d$, so all n points fall in the interval $[t, 1]$ thus $R - L < d$. Therefore, by the law of total probability

$$\begin{aligned}\mathbb{P}(R - L \leq d) &= \int_0^{1-d} f_L(t) \left(\frac{d}{1-t}\right)^{n-1} dt + \mathbb{P}(L \geq 1 - d)\\ &= \int_0^{1-d} n(1-t)^{n-1} \left(\frac{d}{1-t}\right)^{n-1} dt + d^n\\ &= n(1-d)\,d^{n-1} + d^n.\end{aligned}$$

As a consequence we see that the density function of $R - L$ is given by $f_{R-L}(t) = n(n-1)t^{n-2}(1-t)$. Recall that the density of a Beta distribution is given by $f(t) = \frac{1}{\beta(r,s)} t^{r-1}(1-t)^{s-1}$, where $\beta(r,s) = \frac{\Gamma(r)\Gamma(s)}{\Gamma(r+s)}$. Notice that $\beta(n-1, 2) = \frac{1}{n(n-1)}$, so $f_{R-L}(t) = \frac{1}{\beta(n-1,2)} t^{n-2}(1-t)$ is a Beta density function, as claimed.

Now we prove (15). Take $t \in (0, 1)$. Clearly $L' \leq L$ and $R \leq R'$ so that $R - L \leq R' - L'$, then it follows $\mathbb{P}(R - L \geq t) \leq \mathbb{P}\left(R' - L' \geq t\right)$. Also, $a\rho_i + (1 - 2a)\rho_i \leq ar + (1 - 2a)r$ so $R' - L' \leq R - L + (1 - a)r$, thus $\mathbb{P}\left(R' - L' \geq t\right) \leq \mathbb{P}(R - L + (1 - a)r \geq t)$. Hence

$$\mathbb{P}(R - L \geq t) \leq \mathbb{P}\left(R' - L' \geq t\right) \leq \mathbb{P}(R - L + (1 - a)r \geq t). \tag{16}$$

The law of total probability yields

$$\mathbb{P}\left(R' - L' \leq t\right) = \mathbb{P}\left(R' - L' \leq t \mid G\right)\mathbb{P}(G) + \mathbb{P}\left(R' - L' \leq t \mid G^c\right)\mathbb{P}\left(G^c\right)$$

which implies $\mathbb{P}\left(R' - L' \leq t\right) \leq \mathbb{P}\left(R' - L' \leq t \mid G\right)\mathbb{P}(G) + \mathbb{P}(G^c)$ and $\mathbb{P}\left(R' - L' \leq t\right) \geq \mathbb{P}\left(R' - L' \leq t \mid G\right)\mathbb{P}(G)$. We then obtain the inequality

$$\frac{\mathbb{P}\left(R' - L' \leq t\right) - \mathbb{P}(G^c)}{\mathbb{P}(G)} \leq \mathbb{P}\left(R' - L' \leq t \mid G\right) \leq \frac{\mathbb{P}\left(R' - L' \leq t\right)}{\mathbb{P}(G)} \tag{17}$$

Now we have for $\epsilon \in (0, 1)$, sufficiently large n and for M and c as in Theorem 2 the following

$$\begin{aligned}\mathbb{P}\left(\{R'(n) - L'(n) \leq \epsilon\} \mid G(n)\right) &\leq \frac{\mathbb{P}\left(\{R'(n) - L'(n) \leq \epsilon\}\right)}{\mathbb{P}(G(n))}\\ &\leq \frac{\mathbb{P}\left(\{R'(n) - L'(n) \leq \epsilon\}\right)}{1 - Me^{-c}}\\ &\leq \frac{\mathbb{P}(\{R(n) - L(n) \leq \epsilon\})}{1 - Me^{-c}},\end{aligned}$$

where the first inequality follows from (17), the second from Theorem 2 and the last from $R - L \leq R' - L'$. Now the result follows from the $Beta(n-1, 2)$ distribution of $R(n) - L(n)$. □

4 Simulations

In this section we present the results of simulation experiments illustrating the main findings of Theorems 2, 3, and 4. In Fig. 2, panel (a), we see simulations of random interval graphs in which r follows the regime of Theorem 2 in the form $r(n) = \beta\frac{\log(n)+c}{n}$ for $a = 0.2$, $\beta = \frac{2}{1-a}$ and $c = 3.1$. The number of nodes (n) changes as indicated in each graph in the panel. What we see is that all graphs are connected in agreement with the main claim in Theorem 2 of having connected instances with high probability. In Fig. 2, panel (b), we see simulations of random interval graphs in which $r(\cdot)$ follows the dynamic $r(n) = \alpha\beta\frac{\log(n)}{n}$ with $\alpha = 0.5$ and β as before. What we see is that all of the simulated graphs are disconnected and with many isolated nodes, which is to be expected by Theorem 3.

In Fig. 3 we see simulations illustrating Theorem 4. On the left side Fig. 3a we see the histogram after 2e5 simulations of the difference $R(100) - L(100)$, where the random variables $R(n)$ and $L(n)$ are defined in Theorem 4 and the parameters are as in the simulation exercise for Theorem 2, the only difference is that here the number of nodes is fixed. It presents visual evidence of the

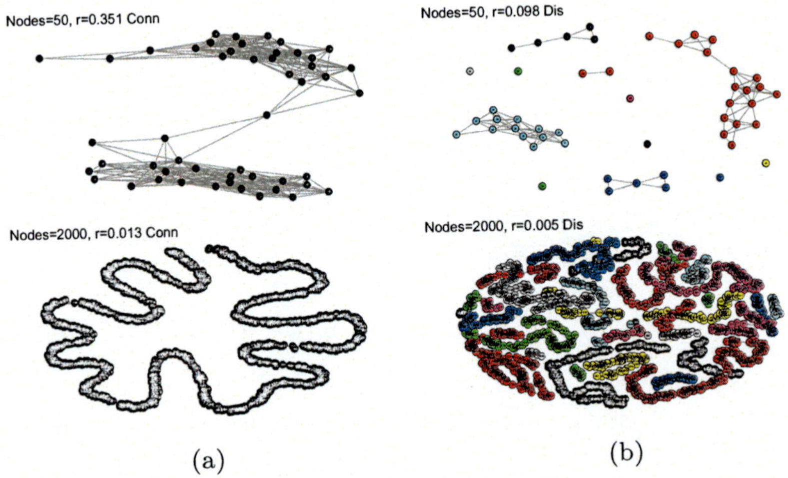

Fig. 2 In panel **(a)** we see simulations of random interval graphs under the specifications of Theorem 2. As expected, graphs are always connected. In panel **(b)** we see simulations of random interval graphs under the specifications of Theorem 3. As expected, graphs have always more than one component

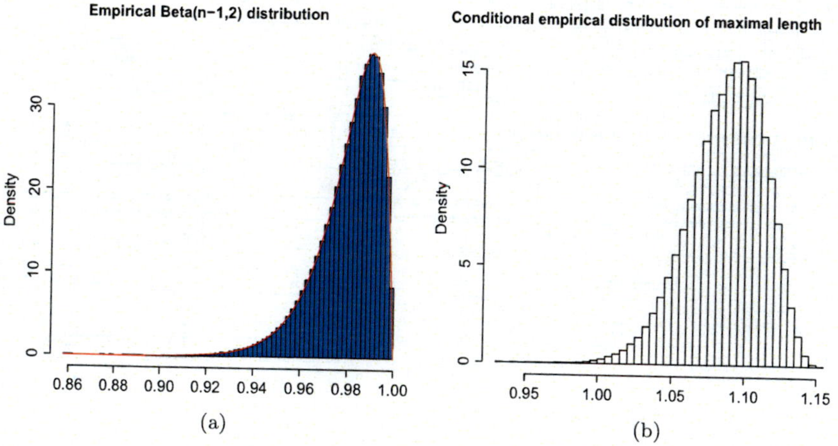

Fig. 3 Empirical distributions illustrating Theorem 4. (**a**) Histogram of $R - L$. The red line displays the density function of a $Beta(n - 1, 2)$ distribution. (**b**) Histogram of $R'(n) - L'(n)$ on the event of a connected graph

convergences to a $Beta(n - 1, 2)$ distribution with $n = 100$. Beyond this evidence, we estimated a Wasserstein distance of 0.00014 between the Beta distribution and the empirical distribution function obtained from the simulations. To this end, we applied R package "transport". On the right side Fig. 3b we see the histogram of $R'(100) - L'(100)$ for the same 2e5 simulations as before but intersected with the event that the interval graph was connected. The random variables $R'(n)$ and $L'(n)$ are defined in Theorem 4. We see a very high probability of reaching or even surpassing the intended length, again, as predicted.

The bifurcation of (non) connectedness for random interval graphs in our asymmetric model has been illustrated on Fig. 2.

In Fig. 3 we started to see the behavior of empirical probabilities and now we go deeper into characterizing by simulation the sensibility of these probabilities with respect to parameters. In all experiments we generated a hundred trials per set of parameters, then we checked connectedness and saved the percentage of connected instances.

In Fig. 4, following the hypothesis of Theorem 2, we fixed two different values for $a = 0.33, 0.66$ in order to numerically illustrate that our requirement of $a < 1/2$ is crucial. We ran simulations for different values of c, while varying the number of agents from 0 to 10,000, to observe how the choice of c affects the convergence speed to obtain connected interval graphs with high probability. For low values of a we found experimentally that a value of c between 3 and 4 suffices, as shown in Fig. 4a. As expected, high values of a yield a slow convergence speed, which is illustrated by the low empirical probabilities of connected interval graphs observed in Fig. 4b.

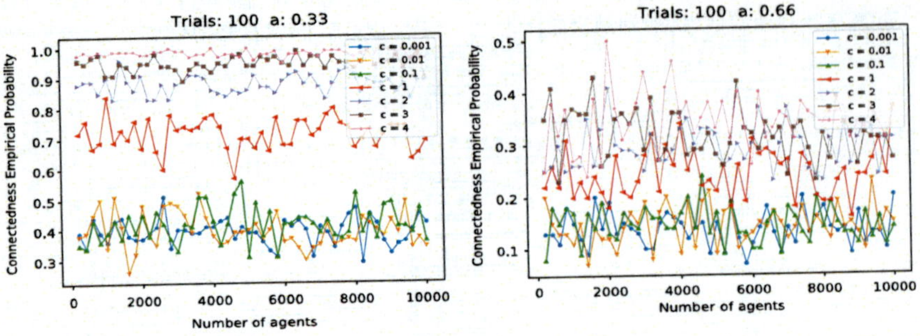

Fig. 4 Simulations illustrating the sensibility of empirical probabilities to parameters. The dynamic of $r(n)$ is according to Theorem 2

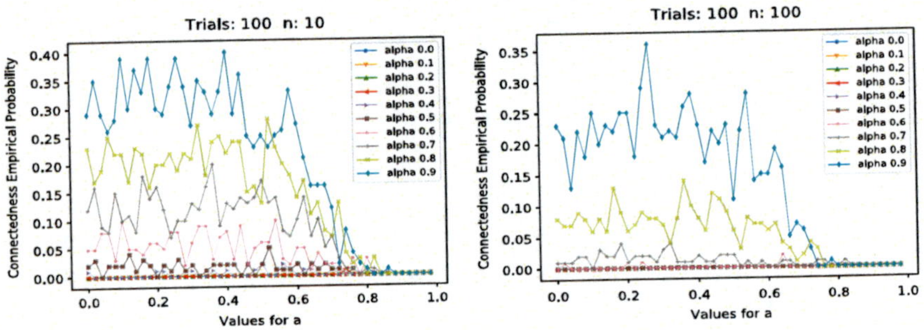

Fig. 5 Simulations illustrating the sensibility of empirical probabilities to parameters. The dynamic of $r(n)$ is according to Theorem 3

Next, following the hypothesis of Theorem 3, we considered two fixed values for n, namely $n = 10$ and 100, to see how the change of the order of magnitude of n affects the approximation of the probability that graph instances are not connected, as stated in Theorem 3, but not only that, we also vary a to observe how fast this result is achieved. Figure 5 shows ten different trajectories for ten values of α. As expected, there is a monotone relationship with α. Experimentally, we found that a value around 0.5 is a critical value where disconnected interval graphs appear with high probability for moderate values of n. As it is to be expected, by Theorem 3, as n grows arbitrarily large, the probability of generating disconnected interval graphs increases for any value of α. Also, experimentally we found that large values of a dramatically increases the probability of generating disconnected interval graphs.

In Fig. 6, following the hypothesis of Theorem 4, additionally to connectedness, we also account for $R' - L' \geq 1$, guaranteeing this way solvability. As before, we compute the empirical probability by executing a hundred trials per set of parameters. We fix the number of agents in $n = 10, 100$, and consider seven different values for c. We observe from the experiments, as a variates from 0 to 1, the empirical probability raises until a peak is reached when $a \approx 3$, consistently for the two values of n. Also, we note that as the value of n increases, the empirical

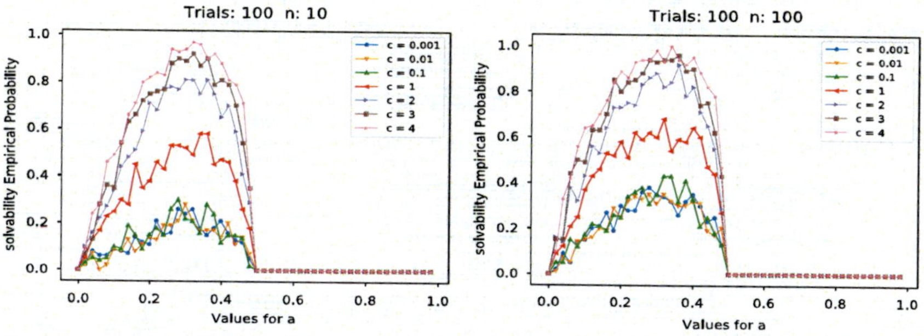

Fig. 6 Simulations illustrating the sensibility of empirical probabilities to parameters. Probabilities are conditioned to the event of hitting the boundaries of the interval [0, 1]. According to Theorem 4, this probability is asymptotically high

probabilities increase for all practical purposes. A last observation is, confirming the hypothesis that $a < \frac{1}{2}$, beyond the value $a \geq \frac{1}{2}$ it is impossible that $R' - L' \geq 1$ and therefore the empirical probability crushes to zero.

5 Conclusion

As we have seen in Theorem 1, the DDLP has an equivalent representation as a graph theoretical problem. Specially, solvability of a DDLP instance translates into the existence of a connected interval graph. This interpretation is far reaching since DDLP is NP-complete, and therefore it is likely that no efficient algorithm can solve it. In contrast, checking for connectedness can be performed very efficiently.

The use of probability theory allowed us to explore the characteristics of a subset of solvable DDLP instances. On the one hand, Theorem 2, showed a threshold for the regime of $r(n)$, under which $G_A(Q, a)$ is connected almost everywhere. Our simulations gave a visual illustration of Theorem 2, where the selection parameter of a and c played an important role in order to observe convergence speed, so that connectedness shows up for reasonable values of n. On the other hand, characterizing disconnectedness in Theorem 3, the function $r(n)$ follows a different regime under which disconnectedness of $G_A(Q, a)$ has high probability.

Disconnectedness in the asymmetric model does not tell us if an instance is not solvable, but the symmetric model does guarantee that a disconnected interval graph renders unsolvable DDLP instances. This will produce fringes for solvability/connectedness and unsolvability/disconnectedness as we now discuss. Indeed, Theorem 2 implies connectedness with high probability under the regime $r(n) = \frac{2}{1-a} \frac{\log(n)+c}{n}$ while Sheinerman shows that for $r(n) \leq \log(n)/n$ almost all graphs $G_S(Q)$ are disconnected, and thus the corresponding DDLP is unsolvable.

Theorem 3 determines that regimes $r(n) = \frac{2}{1-a} \frac{\log(n)\alpha(n)}{n}$ with $\lim_n \alpha(n) < 1$ have high probability of disconnectedness (in a severe way!). Note that there is a gap in between both regimes in which solvability is ambiguous since it depends on asymmetry of the model.

Our simulations for Theorem 3 showed sensitivity to the choice of parameters α and a, in order to appreciate the asymptotic result for reasonable values of n. Lastly, Theorem 4 provides insight into how solvable DDLP instances arise with high probability, as we considered the connectedness of $G_A\left(Q', a\right)$. Our experimentation showed that $a \approx \frac{1}{3}$ and a value of $c > 3$ are good parameters to generate solvable DDLP instances with high probability for most values of n.

The analysis in the paper sets up for a taxonomy to generate feasible-to-solve DDLP instances, which lead to use efficient approaches, such as checking connectedness in an interval graph, as its inherent difficulty makes very difficult to solve DDLP instances in general. The DDLP representation as a graph theory problem, showed to be fruitful, because it allowed us to discover asymptotic behavior in randomly generated instances, analogous to the one described by Erdős and Rényi, and Scheinermann.

References

1. Anaya, J., Chalopin, J., Czyzowics, J., Labourel, A., Pelc, A., Vaxes, Y.: Collecting information by power-aware mobile agents. In: Aguilera, M.K. (ed.) Distributed Computing, volume 7611 of Lecture Notes in Computer Science, Salvador. International Symposium on Distributed Computing. Springer (2012)
2. Arora, S., Barak, B.: Computational Complexity. Cambridge University Press, Cambridge (2009)
3. Bampas, E., Das, S., Derenjowski, D., Karousatou, C.: Collaborative delivery by energy-sharing low-power mobile robots. In: Fernández, A., Jurdzinski, T., Mosteiro, M., Zhang, Y. (eds.) Algorithms for Sensor Systems, volume 10718 of Lecture Notes in Computer Science. ALGOSENSORS. Springer, Cham (2017)
4. Bärtschi, A., Graf, D., Mihalák, M.: Collective fast delivery by energy-efficient agents. In: Potapov, I., Spirakis, P., Worrell, J. (eds.) 43rd International Symposium on Mathematical Foundations of Computer Science, vol. 117, pp. 56:1–56:16. MFCS, LIPICS (2018)
5. Biau, G., Devroye, L.: Lectures on the Nearest Neighbor Method. Springer, Berlin (2015)
6. Bollobás, B.: Random Graphs. Cambridge Studies in Advanced Mathematics. Cambridge University Press, Cambridge (2001)
7. Bondy, J.A., Murty, U.S.R.: Graph Theory. Graduate Texts in Mathematics. Springer, Berlin (2008)
8. Chalopin, J., Das, S., Mihalák, M., Penna, P., Widmayer, P.: Data delivery by energy-constrained mobile agents. In: 9th International Symposium on Algorithms and Experiments for Sensor Systems, Wireless Networks and Distributed Robotics, pp. 111–122. ALGOSENSORS (2013)
9. Chalopin, J., Jacob, R., Mihalák, M., Widmayer, P.: Data delivery by energy-constrained mobile agents on a line. In: Esparza, J., Fraigniaud, P., Husfeldt, T., Koutsoupias, E. (eds.) Automata, Languages, and Programming, volume 8573 of Lecture Notes in Computer Science, pp. 423–434. ICALP, Springer, Berlin, Heidelberg (2014)
10. Cohen, J.E.: The asymptotic probability that a random graph is a unit interval graph, indifference graph, or proper interval graph. Discrete Math. **40**, 21–24 (1982)
11. Cook, S.: The complexity of theorem-proving procedures. In: Proceedings Third Annual ACM Symposium (1971)

12. Czyzowics, J., Krzysztof, D., Moussi, J., Wojciech, R.: Algorithms for Communication Problems for Mobile Agents Exchanging Energy. ArXiv:1511.05987v1 [cs.DS], November (2015)
13. Edmonds, J.: Paths, trees and flowers. Can. J. Math. **17**, 449–467 (1965)
14. Erdős, P., Rényi, A.: On the evolution of random graphs. Publ. Math. Inst. Hung. Acad. Sci. **5**, 17–60 (1960)
15. Garey, M., Johnson, D.: Computers and Intractability. Bell Telephone Laboratories Incorporated, Holmdel (1979)
16. Gilbert, E.: Random graphs. Ann. Math. Stat. **30**(4), 1141–1144 (1959)
17. Karp, R.: Reducibility among combinatorial problems. Complex. Comput. Comput. **40**, 85–103 (1972)
18. Scheinerman, E.: An evolution of interval graphs. Discrete Math. **82**, 287–302 (1990)

Geometric Random Graphs on Circles

Omer Angel and Yinon Spinka

Abstract Given a dense countable set in a metric space, the infinite random geometric graph is the random graph with the given vertex set and where any two points at distance less than 1 are connected, independently, with some fixed probability. It has been observed by Bonato and Janssen that in some, but not all, such settings, the resulting graph does not depend on the random choices, in the sense that it is almost surely isomorphic to a fixed graph. While this notion makes sense in the general context of metric spaces, previous work has been restricted to sets in Banach spaces. We study the case when the underlying metric space is a circle of circumference L, and find a surprising dependency of behaviour on the rationality of L.

Keywords Rado graph · Graph isomorphism · Geometric random graphs

1 Introduction and Main Results

Erdős and Rényi initiated the systematic study of the random graph on n vertices, in which any two vertices are connected with probability p, independently. In 1964, Erdős and Rényi [8] showed that the infinite version of this random graph, where there are countably many vertices and any two are independently connected with probability p, is very different from its finite counterpart. Specifically, there exists a fixed graph R such that the infinite random graph is almost surely isomorphic to R. Moreover, R is the same for all $p \in (0, 1)$. Rado [9] gave a concrete and concise description of R. The graph R (or, more precisely, its isomorphism type) is therefore sometimes called the **Rado graph**. The Rado graph has several nice properties. One such property, which in fact characterizes the graph, is that it is **existentially closed**: for any disjoint finite sets of vertices A and B, there exists a

O. Angel (✉) · Y. Spinka
Department of Mathematics, University of British Columbia, Vancouver, BC, Canada
e-mail: angel@math.ubc.ca; yinon@math.ubc.ca

D. Hernández-Hernández et al. (eds.), *Advances in Probability and Mathematical Statistics*, Progress in Probability 79, https://doi.org/10.1007/978-3-030-85325-9_2

vertex which is adjacent to every vertex in A and to no vertex in B. We refer the reader to [7] for more information on the Rado graph and its properties.

The Erdős–Rényi random graph, both its finite version and its infinite version, are non-geometric models—they are random subgraphs of a complete graph. Random geometric graphs have been studied extensively. In these graphs, the vertices are embedded in some previously defined metric space (X, d), and the probability of a connection depends on the distance between the vertices. If the set of vertices is locally finite, the structure of the resulting graph can be expected to mirror that of the underlying metric space X. However, if the set is dense in X, a very different story unfolds. Bonato and Janssen [2] initiated the study of random geometric graphs in which any two points of a countable metric space that are at distance less than one from each other are independently connected with probability p. They introduced a property of these graphs called **geometric existentially closed** (g.e.c.), analogous to the existentially closed property of the Rado graph. A graph G whose vertex set is a metric space (V, d) is said to satisfy g.e.c. if, for any vertex $s \in V$, any disjoint finite sets of vertices $A, B \subset V$ which are contained in the open unit-ball around s, and any $\epsilon > 0$, there exists a vertex $v \in V \setminus (A \cup B)$ which is adjacent to every vertex in A, is not adjacent to any vertex in B, and satisfies that $d(v, s) < \epsilon$. They then showed that, for any countable metric space in which every point is an accumulation point, the corresponding geometric random graph almost surely satisfies the g.e.c. property.

A countable metric space is said to be **Rado** if the infinite random geometric graph has a unique isomorphism type, i.e., if two independent samples of the geometric random graph, possibly with distinct p, are almost surely isomorphic. Such a space is called **strongly non-Rado** if two such samples are almost surely non-isomorphic. When referring to these terms in the context of a countable subset of a metric space, we are actually referring to the metric space induced on that set. Thus, if S is a countable subset of a metric space (V, d), then we say that S is Rado (strongly non-Rado) if the metric space induced on S, namely $(S, d|_{S\times S})$, is Rado (strongly non-Rado). We informally say that a metric space has the **Rado property** if a typical (e.g., generic or random) dense countable subset of it is a Rado set. To make this precise, if the metric space has some probability measure, one can consider the set S given by an infinite i.i.d. sequence of samples from the measure. The basic question then arises: which metric spaces have the Rado property? For example, if the space has diameter less than 1, then any two points are connected with probability p and the geometric random graph is nothing but the original Rado graph.

In the case of the metric space $(\mathbb{R}, |\cdot|)$, Bonato and Janssen [2] prove that the Rado property holds: there exists a fixed graph, denoted $GR(\mathbb{R})$, such that for a generic dense countable subset of $\mathbb{R}$ (more precisely, for any dense countable set having no two points at an integer distance apart) and for any p, the random graph is almost surely isomorphic to $GR(\mathbb{R})$. They also extend this to the case of the metric space $(\mathbb{R}^d, \ell_\infty)$. Here too, there is a fixed graph $GR(\mathbb{R}^d)$ for each $d \geq 1$. The graphs $R, GR(\mathbb{R}), GR(\mathbb{R}^2), \dots$ are all non-isomorphic to one another. In contrast, they show that this is not true for the Euclidean metric space $(\mathbb{R}^d, \ell_2)$, where every dense countable set is strongly non-Rado [2], nor for the hexagonal norm on $\mathbb{R}^2$, where a randomly chosen dense countable set is almost surely strongly non-Rado [3]. They later showed that many normed spaces fail to have the Rado property [4], including $(\mathbb{R}^d, \ell_p)$ for any $1 < p < \infty$ (and also for $p = 1$ when $d \geq 3$). Balister, Bollobás, Gunderson, Leader and Walters [1] subsequently showed that $(\mathbb{R}^d, \ell_\infty)$ are the unique finite-dimensional normed spaces for which the Rado property holds. In fact, in any other normed space, a generic dense countable subset is strongly non-Rado.

Certain infinite-dimensional normed spaces (all of which are Banach spaces) have also been considered. Bonato, Janssen and Quas [5] studied the space $\mathfrak{c}$ of real convergent sequences with the sup norm and the subspace $\mathfrak{c}_0$ of sequences converging to zero, and showed that both have the Rado property. They also showed that Banach spaces can be recovered from the random graph in the sense that, if two Banach spaces yield isomorphic random graphs, then the two spaces are isometrically isomorphic. In a subsequent paper [6], the same authors considered the space $C[0, 1]$ of continuous functions with sup norm, and proved that the Rado property holds for certain subspaces, including the spaces of piece-wise linear paths, polynomials, and Brownian motion paths.

Though the notion of an infinite random geometric graph is defined in the general context of (countable) metric spaces, we are unaware of any previous works which deal with metric spaces other than normed spaces. One of the goals of this paper is to investigate the random graph in such spaces, and we do so through the example of the cycle.

Let $L > 0$ and consider $\mathbb{S}_L := \mathbb{R}/L\mathbb{Z}$, the circle of circumference L with its intrinsic metric (so that, for example, the diameter of the metric space is $L/2$). Let S be a dense countable subset of $\mathbb{S}_L$. Let $G_{L,S}$ be the unit-distance graph on S, i.e., the graph whose vertex set is S and whose edge set consists of all pairs of points in S whose distance in $\mathbb{S}_L$ is less than 1. Given $p \in (0, 1)$, let $G_{L,S,p}$ be a random subgraph of $G_{L,S}$ obtained by retaining each edge of $G_{L,S}$ with probability p, independently for different edges. See Fig. 1 for an example with a finite set S.

As usual, we say that two graphs G and G' are isomorphic if there exists a bijection φ from the vertex set of G to that of G' such that $\varphi(u)$ and $\varphi(v)$ are adjacent in G' if and only if u and v are adjacent in G. In this case, we write $G \cong G'$. If $L \leq 2$, then $G_{L,S,p}$ is easily seen to be isomorphic to the Rado graph R. Thus, we henceforth always assume that $L \geq 2$.

Our first result is concerned with distinguishing between the different metric spaces, showing that different values of L produce non-isomorphic graphs, so that one can recover the length L of the circle from (the isomorphism type of) the random graph. When $L = \infty$, it is natural to interpret $\mathbb{S}_\infty$ as the metric space $(\mathbb{R}, |\cdot|)$.

Theorem 1.1 *For any $L \in [2, \infty]$, any dense countable $S \subset \mathbb{S}_L$ and any $p \in (0, 1)$, the cycle length L can be recovered almost surely as a measurable function of the graph $G_{L,S,p}$ which is invariant to graph isomorphisms.*

Remark 1.2 There is a minor delicate issue with the definition of recoverability of L. For a graph on some countable set of vertices, which may be enumerated by $\mathbb{N}$, we have the usual product σ-algebra generated by the presence of edges. The claim is that there exists a measurable function f from the set of graphs on vertex set $\mathbb{N}$

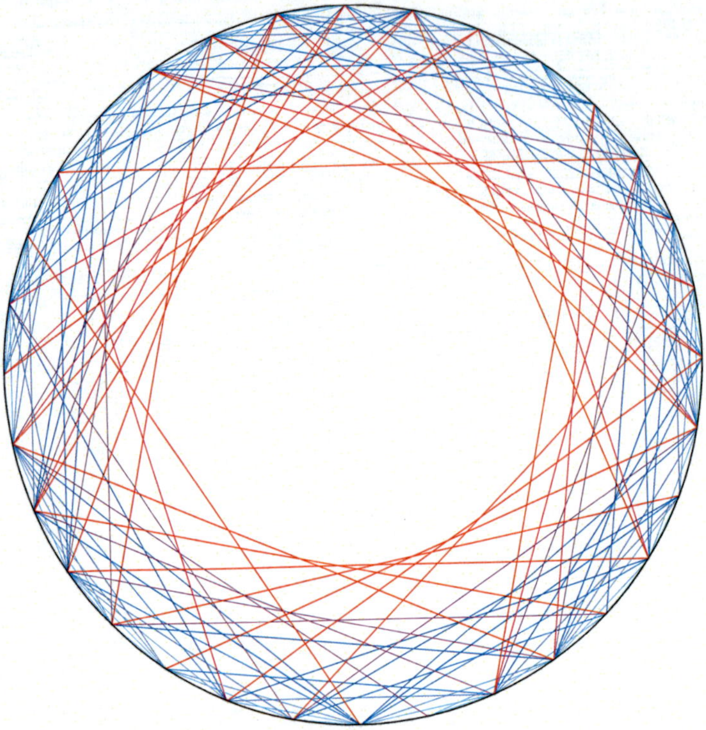

Fig. 1 A random geometric graph in $\mathbb{S}_3$ with 32 equally spaced vertices

to $\mathbb{R}_+$ such that, for any $L \in [2, \infty]$, any $p \in (0, 1)$, any dense countable $S \subset \mathbb{S}_L$ and any enumeration of S, we almost surely have $f(G_{L,S,p}) = L$. Moreover, f is invariant to relabeling the vertices of the graph. Crucially, this invariance is complete and not just probabilistic. That is, $f(G)$ is invariant to any permutation of the vertex labels of G, even if the permutation is allowed to depend on which edges are present in G. To clarify the strength of this property, consider i.i.d. sequences (X_i) of Bernoulli random variables. The expectation p can be recovered almost surely from the law of large numbers as $\lim \frac{1}{n} \sum_{i \le n} X_i$. However, this function is not invariant to arbitrary permutations of the sequence if the permutation is allowed to depend on the sequence. The reason our function is strongly invariant to relabeling is that, for any ℓ, the set of graphs with $f(G) \le \ell$ is described as those graphs satisfying a certain second-order proposition, which does not involve the labels.

Our second and main result is concerned with self-distinguishability, showing that $\mathbb{S}_L$ has the Rado property if and only if L is rational. For rational L, we say that a set $S \subset \mathbb{S}_L$ is **integer-distance-free** if no two points in S are at a distance which is a multiple of $\frac{1}{m}$, where $L = \frac{\ell}{m}$ is irreducible. This terminology may seem mysterious, so we remark that a set S is integer-distance-free if and only if, starting at any point of S and moving an integer distance along the cycle, possibly winding multiple times around the cycle, one can never terminate at another point of S.

Theorem 1.3 *Let $L \geq 2$, let S, S' be dense countable subsets of $\mathbb{S}_L$ and let $p, p' \in (0, 1)$. Let $G = G_{L,S,p}$ and $G' = G_{L,S',p'}$ be independent. Then, almost surely,*

1. *$G \ncong G'$ if $L \notin \mathbb{Q}$.*
2. *$G \cong G'$ if $L \in \mathbb{Q}$ and S and S' are integer-distance-free.*

The theorem implies that, for rational L, a generic dense countable set is a Rado set, whereas, for irrational L, every such set is strongly non-Rado. In the rational case, there exist dense countable sets which are non-Rado (see Remark 5.4). In the irrational case, we can show more—namely, that up to isometries of $\mathbb{S}_L$, one may recover S from (the isomorphism type of) the random graph.

Theorem 1.4 *Let $L > 2$ be irrational, let S, S' be dense countable subsets of $\mathbb{S}_L$ and let $p, p' \in (0, 1)$. Suppose that $G = G_{L,S,p}$ and $G' = G_{L,S',p'}$ can be coupled so that $G \cong G'$ with positive probability. Then S and S' differ by an isometry of $\mathbb{S}_L$.*

2 Definitions and Notation

We view elements of $\mathbb{S}_L$ as real numbers modulo L, and we sometimes identify $\mathbb{S}_L$ with the interval $[0, L)$. An **open arc** is the image of an open interval in $\mathbb{R}$ modulo L. For $a, b \in \mathbb{S}_L$, we write $(a, b)_{\mathbb{S}_L}$ for the positive/anti-clockwise open arc from a to b, i.e., for (a, b) when $0 \leq a \leq b < L$ and for $(a, b + L)$ modulo L when $0 \leq b < a < L$. Thus, $(a, b)_{\mathbb{S}_L}$ and $(b, a)_{\mathbb{S}_L}$ partition $\mathbb{S}_L \setminus \{a, b\}$. The length of an open arc $(a, b)_{\mathbb{S}_L}$ is $b - a$ if $b \geq a$ and is $b - a + L$ if $a > b$. When a and b are real numbers (not necessarily in $\mathbb{S}_L$) with $\|a - b\| < L$, we may unambiguously define $(a, b)_{\mathbb{S}_L}$ by interpreting a and b modulo L. With a slight abuse of notation, we simply write (a, b) for $(a, b)_{\mathbb{S}_L}$. Closed arcs and half-open-half-closed arcs are similarly defined. The distance between two points $u, v \in \mathbb{S}_L$, denoted $\|u - v\|$, is the length of the shorter of the two arcs between u to v and can be written as

$$\|u - v\| = \min\{|u - v|, L - |u - v|\}.$$

Given a graph G, we write $N(v)$ for the neighbourhood of a vertex v in G, and we write $\text{dist}_G(u, v)$ for the graph-distance between vertices u and v in G. The length of a path in G is the number of edges in the path, so that $\text{dist}_G(u, v)$ is the length of a shortest path between u and v.

A graph G whose vertex set is a subset S of $\mathbb{S}_L$ is called **g.e.c.** (geometrically existentially closed) if, for any vertex $s \in S$, any disjoint finite sets $A, B \subset S$ which are contained in $(s - 1, s + 1)$, and any $\epsilon > 0$, there exists a vertex $v \in S \setminus (A \cup B)$ which is adjacent to every vertex in A, is not adjacent to any vertex in B, and satisfies that $\|v - s\| < \epsilon$. The graph G is said to have **unit threshold** if any two adjacent vertices $u, v \in S$ satisfy that $\|u - v\| < 1$. The notions of g.e.c. and unit threshold exist in the general context of a graph whose vertex set is a metric space.

For a dense countable subset S of $\mathbb{S}_L$, let $\mathcal{G}_{L,S}$ denote the collection of all g.e.c. graphs on S having unit threshold. Let $\mathcal{G}_L$ denote the union of $\mathcal{G}_{L,S}$ over all such S. As it turns out, once we establish the following simple property that $G_{L,S,p}$ belongs to $\mathcal{G}_{L,S}$, almost all our arguments become completely deterministic.

Lemma 2.1 *Let $L \in [2, \infty]$, let S be a dense countable subset of $\mathbb{S}_L$ and let $p \in (0, 1)$. Then, almost surely, $G_{L,S,p} \in \mathcal{G}_{L,S}$.*

Proof It is trivial from the definition that $G_{L,S,p}$ almost surely has unit threshold. Fix a vertex $s \in S$, disjoint finite sets $A, B \subset S$ which are contained in the open unit-ball around s, and a rational $0 < \epsilon < 1 - \max_{u \in A \cup B} \|u - s\|$. Since S is countable, it suffices to show that, almost surely, there exists a vertex $v \notin A \cup B$ which is adjacent to every vertex in A, is not adjacent to any vertex in B, and satisfies that $\|v - s\| < \epsilon$. Since the open ball of radius ϵ around s contains infinitely many points v which are not in $A \cup B$, and since the sets of edges incident to these v are independent, it suffices to show that each such v has a positive (fixed) probability to be adjacent to all of A and to none of B. Indeed, since any such v has $\|u - v\| < 1$ for all $u \in A \cup B$, this event has probability $p^{|A|}(1 - p)^{|B|}$. □

3 Distinguishing Graphs Arising from Different L

In this section, we prove Theorem 1.1. Our strategy is to introduce a graph-theoretic quantity which allows to differentiate between graphs arising from different L.

For a graph G, define $\lambda(G)$ to be the supremum over $\lambda \geq 0$ such that for every finite set of vertices $U \subset G$, there exists a vertex in G having at least $\lambda|U|$ neighbours in U. Thus,

$$\lambda(G) := \inf_{\substack{U \subset G \\ 0<|U|<\infty}} \sup_{v \in G} \frac{|N(v) \cap U|}{|U|}.$$

Consider a graph $G \in \mathcal{G}_{L,S}$. It is easy to check that $\lambda(G) = 1$ if $L \leq 2$, since G is just the Rado graph. If $L = \infty$, then $\lambda(G) = 0$, since S contains arbitrarily large finite sets U such that all vertices of U are at distance more than 2 from each other, so that $|N(v) \cap U| \leq 1$ for all v. In fact, as we now show, $\lambda(G)$ depends on G only through L, and moreover, is equal to $2/L$. Theorem 1.1 is an immediate consequence of Lemma 2.1 and the following.

Proposition 3.1 *Let $L \in [2, \infty]$ and $G \in \mathcal{G}_L$. Then $\lambda(G) = 2/L$.*

Proof Let $L \in [2, \infty]$ and $G \in \mathcal{G}_L$. By definition of $\mathcal{G}_L$, the vertex set of G is a dense countable subset S of $\mathbb{S}_L$. For a finite $U \subset S$ and an arc $A \subset \mathbb{S}_L$, we call $|U \cap A|/|U|$ the *density of U in A*.

We begin by proving the upper bound $\lambda(G) \leq 2/L$. Since G has unit threshold, for any v, $N(v)$ is contained in an arc of length 2. Thus it suffices to exhibit, for any $\epsilon > 0$, a finite set $U \subset S$ whose density is no more than $2/L + \epsilon$ in any arc of length 2. Any set that is close to evenly distributed on the cycle will do. For completeness, here is one construction: Let n be a large integer and consider the set V consisting of points $v_0, \ldots, v_{n-1} \in \mathbb{S}_L$, where $v_i := iL/n$. Since S is dense, there exist a finite set U consisting of points $u_0, \ldots, u_{n-1} \in S$ such that $\|u_i - v_i\| < 1/n$. It is straightforward to verify that, for any arc A of length r, we have that $|U \cap A| \leq |V \cap A| + 2 \leq \lfloor rn/L \rfloor + 3$. Thus, U has density at most $2/L + 3/n$ in any arc of length 2.

We now turn to the lower bound $\lambda(G) \geq 2/L$. To show this, we show that the situation described above is essentially the worst case. Precisely, given a finite $U \subset S$, we claim that there exists an open arc A of length 2 in which U has density at least $2/L$. This is easy to verify, since if x is uniform in $\mathbb{S}_L$, then the expected number of points of U in the arc $(x-1, x+1)$ is $(2/L)|U|$, so for some x it is at least that large. Since S is dense, and G is g.e.c., the arc A contains a vertex $v \in S$ which is adjacent to all vertices of U in A. This proves the lower bound $\lambda(G) \geq 2/L$. □

4 Recovering Distances and Non-isomorphism for Irrational L

In this section, we prove part (1) of Theorem 1.3, namely that for irrational $L > 2$, the independent graphs $G_{L,S,p}$ and $G_{L,S',p'}$ are almost surely non-isomorphic. The key step is to show that by looking at the graph-structure of G alone, it is possible to determine the distance in $\mathbb{S}_L$ between any two vertices.

Throughout this section, we fix $L \in [2, \infty]$ and assume $G \in \mathcal{G}_{L,S}$ for some dense countable $S \subset \mathbb{S}_L$. It is easy to check that, for any two vertices $u, v \in S$ and any integer $k \geq 2$, we have

$$\|u - v\| < k \qquad \Longleftrightarrow \qquad \mathrm{dist}_G(u, v) \leq k.$$

Indeed, if $\|u - v\| < k$, then for some $\varepsilon > 0$, there is a path $u = x_0, x_1, \ldots, x_k = v$ with $\|x_i - x_{i-1}\| < 1 - \varepsilon$. By g.e.c. there is a perturbation x'_i of x_i for $i = 1, \ldots, k-1$ so that (x'_i, x'_{i-1}) are edges of G for all $i = 1, \ldots, k$. Conversely, the unit-threshold property shows that a path of length at most k in G from u to v implies that the cycle distance is less than k. Note that for $k = 1$ this equivalence fails, since $\{u, v\}$ may or may not be an edge of G. See [2, Theorem 2.4] for a similar statement.

We may rewrite the above as

$$\mathrm{dist}_G(u, v) = \begin{cases} \lfloor \|u - v\| \rfloor + 1 & \text{if } \|u - v\| \geq 1 \\ 1 \text{ or } 2 & \text{if } \|u - v\| < 1 \end{cases}. \tag{1}$$

Thus, distances in G are predetermined for points of S which are at distance at least 1 in $\mathbb{S}_L$. However, we are more interested in the other direction of implication: forgetting that the vertices of G are labelled by elements of $\mathbb{S}_L$ and looking only at the graph structure of G in relation to (u, v), one may recover $\lfloor \|u - v\| \rfloor$, unless it is 0 or 1.

To formalize these types of ideas, we require some definitions. A **graph with k distinguished vertices** is a graph G, together with an ordered k-tuple $(v_1, \ldots, v_k)$ of distinct vertices of G. Let $\mathcal{G}_{L,\bullet,\bullet}$ denote the collection of all graphs in $\mathcal{G}_L$ with two distinguished vertices. Let π denote the projection from $\mathcal{G}_{L,\bullet,\bullet}$ to the class $\mathcal{G}_{\bullet,\bullet}$ of isomorphism classes of graphs with two distinguished vertices. The above may be restated as saying that the function $(G, u, v) \mapsto \lfloor \|u - v\| \rfloor \mathbf{1}_{\{\|u-v\| \geq 2\}}$ from $\mathcal{G}_{L,\bullet,\bullet}$ to $\mathbb{Z}$ can be written as a composition of a function from $\mathcal{G}_{\bullet,\bullet}$ to $\mathbb{Z}$ with π. Indeed, (1) gives a concrete such description, since the graph-distance between the distinguished vertices is invariant to isomorphisms of the graph. In this case, we say that $\lfloor \|u - v\| \rfloor \mathbf{1}_{\{\|u-v\| \geq 2\}}$ can be recovered from the graph structure of (G, u, v).

More generally, we say that a function $f : \mathcal{G}_{L,\bullet,\bullet} \to \Omega$ **can be recovered** from the graph structure of (G, u, v) if $f = F \circ \pi$ for some $F : \mathcal{G}_{\bullet,\bullet} \to \Omega$. For brevity, we say that $f(G, u, v)$ can be recovered from $\pi(G, u, v)$. We extend these definitions to graphs with k distinguished points, writing π also for the projection from $\mathcal{G}_{L,\bullet,\ldots,\bullet}$ to the corresponding set of isomorphism classes of graphs with k distinguished vertices. We shall also talk about sets of vertices being recoverable from the graph. For example, Lemma 4.4 below says that the set of vertices in the shorter arc between u and v along the cycle is recoverable. Formally, a set $A = A(G, u, v)$ of vertices of G can be recovered from $\pi(G, u, v)$, if the function $\mathbf{1}_{\{x \in A\}}$ can be recovered from $\pi(G, u, v, x)$ for any $x \in G$.

The main ingredient in this section is the following proposition, which shows that we can recover plenty of information on the distances in $\mathbb{S}_L$ from the graph structure (for both rational and irrational L). Given this, part (1) of Theorem 1.3 is easily deduced.

Proposition 4.1 *Let $L > 2$, let $G \in \mathcal{G}_L$ and let $u, v \in G$ be adjacent. Then the sequence*

$$(\lfloor \|u - v\| + kL \rfloor)_{k \geq 1}$$

can be recovered from $\pi(G, u, v)$.

The values $\|u - v\| + kL$ can be thought of as the distance from u to v, moving k additional times around the cycle instead of directly. The assumption that u, v are adjacent in G can be removed from this proposition, but it simplifies the proof and does not significantly impact the application for the proof of Theorem 1.3. In the case of irrational L, this gives the following, stronger corollary, which immediately implies the more general result.

Corollary 4.2 *Let $L > 2$ be irrational, let $G \in \mathcal{G}_L$ and let $u, v \in G$. Then $\|u - v\|$ can be recovered from $\pi(G, u, v)$.*

Proof Consider first the case when u and v are adjacent in G, so that Proposition 4.1 applies. It suffices to see that the mapping $x \mapsto (\lfloor x + kL \rfloor)_{k \geq 1}$ is injective on $[0, L)$. It is a well known fact that for irrational L the fractional parts $(kL - \lfloor kL \rfloor)$ are dense in $[0, 1]$. Let $0 \leq x < y < L$. Since the fractional parts are dense, it follows that for some k we have $\lfloor x + kL \rfloor \neq \lfloor y + kL \rfloor$, and the sequences differ.

For any path in G, we can therefore recover from G the total length of the edges along $\mathbb{S}_L$. If u and v are not adjacent, then there is a path in G from u to v which moves around $\mathbb{S}_L$ in the shorter direction without backtracking. Since we can recover the cycle distance in each edge of the path, the sum is the distance from u to v. Any other path in G must give a larger sum. Thus we can recover $\|u - v\|$ as the minimal sum of cycle distances along paths from u to v in the graph. □

Since the cycle distance between any two vertices can be recovered from the graph, we have the following.

Corollary 4.3 *Let $L > 2$ be irrational, let S, S' be dense countable subsets of $\mathbb{S}_L$ and $G \in \mathcal{G}_{L,S}$, $G' \in \mathcal{G}_{L,S'}$. If $f : S \to S'$ is a graph isomorphism between G and G', then it is an isometry between S and S'.*

Corollary 4.3 immediately implies Theorem 1.4. We now prove part (1) of Theorem 1.3.

Proof of Theorem 1.3(1) Let $G = G_{L,S,p}$ and $G' = G_{L,S',p'}$ be independent, as in the statement of the theorem. Consider a bijection $f : S \to S'$. By Corollary 4.3, if f is not an isometry between S and S', then it is not an isomorphism between G and G'. Thus it suffices to consider isometries f. There are at most countably many isometries between S and S' (for an arbitrary $v_0 \in S$, there are at most two isometries for any given choice of $f(v_0)$). Since any fixed isometry f is almost surely not an isomorphism between G and G', we conclude that there almost surely does not exist an isomorphism between G and G'. □

4.1 *Proof of Proposition 4.1*

The overall strategy for the proof of Proposition 4.1 is as follows: we define a graph-theoretic notion of a cyclic ordering of vertices. This notion, though defined completely in terms of the graph G, will be such that it guarantees that the corresponding points in $\mathbb{S}_L$ are cyclically ordered as well. This will allow to define another graph-theoretic notion of a uni-directional path in G, which will correspond to a path in $\mathbb{S}_L$ that winds around the circle in a fixed direction. We then show that any uni-directional path in G has a well-defined (again, in terms of the graph) winding number which counts the number of times the corresponding path in $\mathbb{S}_L$ winds around the circle. Finally, using this we deduce that from the graph G, for any two adjacent vertices $u, v \in G$, we may recover the sequence $(\lfloor \|u - v\| + kL \rfloor)_{k \geq 1}$, which is Proposition 4.1.

Fix $L > 2$, a dense countable subset $S \subset \mathbb{S}_L$ and a graph $G \in \mathcal{G}_{L,S}$. For $x, y \in S$ having $\|x - y\| < 1$, let $A_{x,y}$ denote the set of points of S in the shorter arc of $[x, y]$ and $[y, x]$. It is convenient to include the endpoints x, y in the arc.

The starting point of our argument is the following lemma which shows that the shortest arc between two adjacent vertices can in fact be described as a graph-theoretic property. Its proof is postponed to the end of the section.

Lemma 4.4 *Let $a, b \in G$ be adjacent. Then $A_{a,b}$ can be recovered from $\pi(G, a, b)$.*

We say that a triplet (a, b, c) of distinct points in G is **cyclically ordered in $\mathbb{S}_L$** if $A_{a,b} \cap A_{b,c} = \{b\}$. This means that when moving from a to b to c along the cycle (always via the shorter arc), the direction of movement is maintained. We say that a path $p = (v_0, \dots, v_n)$ in G consisting of distinct vertices is a **uni-directional** if the triplet (v_{i-1}, v_i, v_{i+1}) is cyclically ordered in $\mathbb{S}_L$ for each $1 \le i \le n-1$. Thus, by going along the shorter arc between consecutive vertices, a uni-directional path in G may be thought to correspond to a continuous path in $\mathbb{S}_L$ that always winds around the circle in a single direction. In light of Lemma 4.4, this property can be determined from the graph structure of G, so that we may talk about the path being uni-directional in G. The **winding number** of a uni-directional path p **is defined** as the number of complete revolutions its continuous counterpart makes around the cycle—if its total cycle-length is ℓ, this is $\lfloor \ell/L \rfloor$. The winding number can also be calculated as the number of indices $1 \le i \le n-1$ such that $v_0 \in A_{v_i, v_{i+1}}$. Consequently, the winding number of a uni-directional path p can be recovered from $\pi(G, v_0, \dots, v_n)$.

It will also be useful to be able to identify the direction in which a uni-directional path winds around the circle; by this we do not mean the absolute direction (clockwise/anticlockwise), but rather whether it goes from the start point to the end point by starting through the short/long arc between them. This can be done in one of several ways. We choose here a simple definition, which comes at the cost of requiring the start and end points to be at distance less than 1. For $u, v \in S$, a **good path** from u to v is a uni-directional path $p = (x_0, x_1, \dots, x_n)$ in G such that $x_0 = u$, $x_n = v$ and $v \in A_{u,x_1}$. Thus, a good path is required to go towards v in the shorter direction, and overshoot v in its first step. In particular, its winding number is at least 1. Of course, this is only possible if $\|u - v\| < 1$. The following shows that good paths exist.

Lemma 4.5 *Let $k \ge 1$ and $u, v \in G$ be such that $\|u - v\| < 1$. Then G contains a good path from u to v with winding number k and length $n = \lfloor \|u - v\| + kL \rfloor + 1$. Moreover, there is no good path from u to v with winding number k and length less than n.*

Proof For concreteness, we assume that the short arc from u to v is in the positive direction, so that $v = u + \|u - v\|$. Set $\ell := \|u - v\|$, so that we seek a path of length $n = \lfloor \ell + kL \rfloor + 1$. We start by specifying approximate locations for the points of the path. These approximate locations, denoted x_i, are points of the circle and need not be vertices of the graph. Let $x_1 = u + t$ for some $t \in (\ell, 1)$, so that $v \in A_{u,x_1}$ and

$\|x_1 - u\| < 1$. The total cycle-length of the path will be $\ell + kL$, and the remaining points $(x_i)_{i=2}^{n-1}$ will be equally spaced with gap

$$\Delta = \frac{\ell + kL - t}{n-1}.$$

Thus, the approximate points are $x_i = u + t + \Delta(i-1)$ for $1 \leq i \leq n-1$. Note that for t close enough to 1, we have $\Delta < 1$, since $\ell + kL < n$.

Fix $\varepsilon > 0$ such that $\max\{t, \Delta\} < 1 - 2\varepsilon$, and set $U_0 := \{u\}$, $U_n := \{v\}$ and $U_i := S \cap (x_i - \varepsilon, x_i + \varepsilon)$ for $1 \leq i \leq n-1$. This choice guarantees that any point in U_i is at distance less than 1 from any point in U_{i-1}, and the shorter arc between them is positively oriented. Since G is g.e.c., there exists $u_1 \in U_1$ such that u_1 is adjacent to $u_0 := u$ in G. Continuing by induction, we see that there exists a sequence $(u_i)_{0 \leq i \leq n-2}$ such that $u_i \in U_i \setminus \{v, u_0, \ldots, u_{i-1}\}$ and u_i is adjacent to u_{i-1} in G. Finally, there exists $u_{n-1} \in U_{n-1} \setminus \{v, u_0, \ldots, u_{n-2}\}$ which is adjacent to both u_{n-2} and $u_n := v$. By construction, $(u_0, \ldots, u_n)$ is a good path in G from u to v of length n and winding number k.

Finally, there can be no uni-directional path (good or otherwise) from u to v with winding number k and length at most $\lfloor \ell + kL \rfloor$, since the total length of the arcs in such a path is smaller than $\ell + kL$. □

We are now ready to prove Proposition 4.1.

Proof of Proposition 4.1 Let $u, v \in G$ be adjacent and fix $k \geq 1$. Our goal is to recover $\ell_k := \lfloor \|u - v\| + kL \rfloor$ from $\pi(G, u, v)$. By Lemma 4.5, $\ell_k + 1$ is the shortest length of a good path from u to v with winding number k. Since whether a path $(x_0, \ldots, x_n)$ is good can be determined from $\pi(G, x_0, \ldots, x_n)$, this shows that ℓ_k can be recovered from $\pi(G, u, v)$. □

4.2 Proof of Lemma 4.4

Let $a, b \in G$ be adjacent. Recall that $A_{a,b}$ is the set of points of S in the shorter arc of $[a, b]$ or $[b, a]$. As a warm-up, the reader may find it instructive to note that, when $L \geq 5$, one may easily recover $A_{a,b}$ as the intersection over $v \in S$ of all intervals $(v-2, v+2) \cap S$ which contain $\{a, b\}$. By (1), each such interval can be recovered as the set of vertices at graph-distance at most 2 from v.

When $L \geq 3$, one may try a similar approach, recovering $A_{a,b}$ as the intersection over $v \in S$ of all intervals $(v-1, v+1) \cap S$ which contain $\{a, b\}$. Indeed, it is not hard to check that this produces the correct set, however, as (1) does not allow to recover intervals of the form $(v-1, v+1) \cap S$, we will need to slightly modify the approach.

The proof is split into two cases according to whether $L \geq 3$ or $2 < L < 3$. The following lemma will be useful in both cases. Say that a set $U \subset G$ is **small** if it is finite and some vertex of G is adjacent to every vertex in U. Say that U is **large** if it is finite and not small, i.e., no vertex of G is adjacent to every vertex in U. The following simple lemma shows that a finite set is small if and only if it is contained in some open arc of length 2. Equivalently, a finite set is large if and only if it leaves no gap of length greater than $L-2$ in its complement.

Lemma 4.6 *Let $U \subset G$ be finite. Then U is small if and only if $U \subset (v-1, v+1)$ for some $v \in G$.*

Proof Suppose first that U is small. By definition, there exists a vertex $w \in G$ that is adjacent to every vertex in U. Since G has unit threshold, $\|u - w\| < 1$ for all $u \in U$. Thus, $U \subset (w-1, w+1)$, as required.

Suppose now that $U \subset (v-1, v+1)$ for some $v \in G$. Since G is g.e.c., there exists a vertex $w \in G$ which is adjacent to all of U. Thus, U is small, as required. □

4.2.1 Proof of Lemma 4.4 when $L \geq 3$

Step 1 *Let (a, b, c, d) be a path in G and suppose that $\{a, b, c, d\}$ is large. Then $A_{b,c}$ can be recovered from $\pi(G, a, b, c, d)$.*

We first show that such a path (a, b, c, d) must be uni-directional. Indeed, if the arcs (a, b) and (b, c) are in opposite directions in $\mathbb{S}_L$, then $\{a, b, c, d\} \subset (c-1, c+1)$, and so the set is small. Similarly, if (b, c) and (c, d) are in opposite directions, then $\{a, b, c, d\} \subset (b-1, b+1)$. This shows that $\{a, b, c, d\}$ are distinct vertices and that $A_{a,b} \cap A_{b,c} = \{b\}$ and $A_{b,c} \cap A_{c,d} = \{c\}$.

For a finite $U \subset G$, we denote

$$C(U) := \{w \in G : U \cup \{w\} \text{ is small}\}.$$

We will show that $A_{a,b} \cup A_{b,c}$ is precisely the set

$$W := \{w \in G : C(\{a, b, c\}) = C(\{a, b, c, w\})\}.$$

Since $A_{b,c} \cup A_{c,d}$ is similarly obtained, this will show that we can determine $A_{b,c}$ as

$$A_{b,c} = (A_{a,b} \cup A_{b,c}) \cap (A_{b,c} \cup A_{c,d}).$$

Before showing that $A_{a,b} \cup A_{b,c} = W$, we first observe that any interval $(v-1, v+1)$ containing $\{a, b, c\}$, contains $A_{a,b} \cup A_{b,c}$ and does not contain d. Indeed, no such interval contains $\{a, b, c, d\}$ since $\{a, b, c, d\}$ is large. Suppose that (a, b, c, d) winds in the positive direction (the other case being similar). Since $L \geq 3$, the circle is a disjoint union of the arcs $[a, b)$, $[b, c)$, $[c, d)$, and $[d, a)$. Thus any interval in the circle that contains a, b, c but not d also contains $[a, b] \cup [b, c]$. Since $[a, b]$ is the short arc between a, b, we have $A_{a,b} = [a, b] \cap S$, and similarly for $A_{b,c}$. Thus any such interval $(v-1, v+1)$ must contain $A_{a,b} \cup A_{b,c}$.

To see that $A_{a,b} \cup A_{b,c} \subset W$, fix $w \in A_{a,b} \cup A_{b,c}$ and let us show that $C(\{a, b, c\}) = C(\{a, b, c, w\})$. The containment $C(\{a, b, c, w\}) \subset C(\{a, b, c\})$ is clear, and the opposite containment follows from the fact that any interval $(v - 1, v + 1)$ containing $\{a, b, c\}$ also contains $A_{a,b} \cup A_{b,c}$.

To see that $W \subset A_{a,b} \cup A_{b,c}$, let $w \in W$ and suppose towards a contradiction that $w \notin A_{a,b} \cup A_{b,c}$. In order to reach a contradiction with the fact that $C(\{a, b, c\}) = C(\{a, b, c, w\})$, it suffices to find a vertex of G which belongs to an interval $(v - 1, v + 1)$ containing $\{a, b, c\}$, but does not belong to any interval $(v - 1, v + 1)$ containing $\{a, b, c, w\}$.

Recall that (a, b, c, d) is a uni-directional path, and note that one of (a, b, c, d, w) or (a, b, c, w, d) is also a uni-directional path. Suppose for concreteness that it is the latter, and that moreover, (a, b, c, w, d) winds around the cycle in the positive direction. In particular, $A_{a,b} = [a, b] \cap S$ and $A_{b,c} = [b, c] \cap S$. Moreover, $A := [c, w] \cap S$ is the arc between c and w, which does not contain a, b or d (it is not necessarily the short arc between c and w). Note that, since any interval $(v - 1, v + 1)$ containing $\{a, b, c\}$ must contain $A_{a,b} \cup A_{b,c}$ and cannot contain d, it follows that any interval $(v - 1, v + 1)$ containing $\{a, b, c, w\}$ must contain $A_{a,b} \cup A_{b,c} \cup A = [a, w] \cap S$. Observe also that any such interval is contained in $(w - 2, a + 2)$. However, if $v \in (c - 1, w - 1)$, then the interval $(v - 1, v + 1)$ contains $\{a, b, c\}$, but is not contained in $(w - 2, a + 2)$ (note that the latter is not all of $\mathbb{S}_L$ since (a, w) is longer than (a, c), which in turn has length more than 1). We have thus reached a contradiction.

Step 2 *Let $a, b \in G$ be adjacent. Then $A_{a,b}$ can be recovered from $\pi(G, a, b)$.*

In light of the previous step, it suffices to show that there exists a path (x, a, b, y) so that $\{x, a, b, y\}$ is large. Denote $\ell := \|a - b\|$ and suppose they are oriented so that $b = a + \ell$. There exists $x \in (a - 1, a - 1 + \ell/2)$ adjacent to a, and similarly, there exists $y \in (b + 1 - \ell/2, b + 1)$ adjacent to b. Since $L \geq 3$, the only way $\{x, a, b, y\}$ is contained in an open arc of length 2 is if $y - x < 2$, which is not the case by our choice of x and y.

4.2.2 Proof of Lemma 4.4 when $2 < L < 3$

We write $L = 2 + \delta$ for some $\delta \in (0, 1)$. The reader may keep in mind that small δ is the more difficult case.

Step 1 *Let $u, v \in G$. Then $\mathbf{1}_{\{\|u-v\|<\delta\}}$ can be recovered from $\pi(G, u, v)$.*

This will follow if we show that

$$\|u - v\| < \delta \quad \Longleftrightarrow \quad U \cup \{u\} \text{ and } U \cup \{v\} \text{ are large for some small } U.$$

Suppose first that $U \cup \{u\}$ and $U \cup \{v\}$ are large for some small U. Let us show that $\|u - v\| < \delta$. Let $\{u^-, u^+\}$ be the two vertices of U nearest to u from either side. Recall that a finite set is large if and only if it leaves no gap of length greater than $\delta = L - 2$ in its complement. Therefore, since $U \cup \{u\}$ is large, (u^-, u) and

(u, u^+) each has arc-length at most δ. Since $U \cup \{v\}$ is large, but U is small, it must be that $v \in (u^-, u^+)$ so that $\|u - v\| < \delta$, as required.

Suppose next that $\|u - v\| < \delta$. Let us show that $U \cup \{u\}$ and $U \cup \{v\}$ are large for some small U. Assume without loss of generality that $u \in (v, v + \delta)$, and let $0 < \varepsilon < (\delta - \|u - v\|)/3$. Let V be the arc $(v + \delta - \varepsilon, v - 2\varepsilon)$. Note that $v \notin V$ and $|V| = L - (\delta + \epsilon) = 2 - \varepsilon < 2$. Recall that a finite set is small if and only if it is contained in some open arc of length 2. Thus, any finite subset of V is small. Since $(v - 2\varepsilon, v)$ and $(v, v + \delta - \varepsilon)$ each has arc-length less than δ, it follows that $U_1 \cup \{v\}$ is large for some finite $U_1 \subset V$. Similarly, since $(v - 2\varepsilon, u)$ and $(u, v + \delta - \varepsilon)$ each has arc-length less than δ, we have that $U_2 \cup \{u\}$ is large for some finite $U_2 \subset V$. Thus, $U := U_1 \cup U_2$ is a small set such that both $U \cup \{v\}$ and $U \cup \{u\}$ are large, as required.

Step 2 *Let $a, b \in G$ satisfy $\|a - b\| < \delta$. Then $A_{a,b}$ can be recovered from $\pi(G, a, b)$.*

By the first step, $A_v := (v - \delta, v + \delta) \cap S$ can be recovered from $\pi(G, v)$. Let W be the intersection of all A_v containing $\{a, b\}$. We claim that $A_{a,b} = W$.

To see that $A_{a,b} \subset W$, we must show that A_v contains $A_{a,b}$ whenever it contains $\{a, b\}$. Indeed, if $\{a, b\} \subset A_v$ then, since A_v has arc-length 2δ, $\|a - b\| < \delta$ and $3\delta < L$, it follows that $A_{a,b} \subset A_v$.

To see that $W \subset A_{a,b}$, we must show that for any $u \in G \setminus A_{a,b}$ there exists $v \in G$ such that $\{a, b\} \subset A_v$ and $u \notin A_v$. Since $2\delta < L$, it is straightforward that such a v exists.

Step 3 *Let $a, b \in G$ be adjacent. Then $A_{a,b}$ can be recovered from $\pi(G, a, b)$.*

Set $n := \lceil 1/\delta \rceil$, so that $1 \le \delta n < 1 + \delta$. Since $\|a - b\| < 1 \le \delta n$, g.e.c. implies that there exists a uni-directional path $P = (u_0, \ldots, u_n)$ in G such that $u_0 = a$, $u_n = b$, and $\|u_i - u_{i-1}\| < \delta$ for all i. Since the long arc from a to b has length larger than $1 + \delta$, and since the total cycle-length of P is at most $\delta n < 1 + \delta$, it must be that $A_{a,b} = A_{u_0,u_1} \cup \cdots \cup A_{u_{n-1},u_n}$. Thus, by the previous steps, one can recover $A_{a,b}$ from $\pi(G, a, b)$.

5 Constructing an Isomorphism for Rational L

In this section, we prove part (2) of Theorem 1.3, namely, that $G_{L,S,p}$ and $G_{L,S',p'}$ are almost surely isomorphic when L is rational and S and S' are integer-distance-free. In light of Lemma 2.1, this is an immediate consequence of the following. Let $\mathcal{G}_{L,\mathrm{idf}} \subset \mathcal{G}_L$ consist of those graphs whose vertex sets are integer-distance-free.

Proposition 5.1 *Let $L > 2$ be rational and let $G, G' \in \mathcal{G}_{L,idf}$. Then $G \cong G'$.*

Bonato and Janssen proved the analogous statement for the case $L = \infty$ (corresponding to the real line) [2, Theorem 3.3]. Our construction follows similar lines.

Let $G \in \mathcal{G}_{L,S}$ and $G' \in \mathcal{G}_{L,S'}$, where S and S' are integer-distance-free. Our goal is to construct an isomorphism f between G and G'. In fact, we have the flexibility to map one vertex of G to an arbitrary vertex of G'. Thus, by translating S and S' if necessary, we may assume without loss of generality that both S and S' contain 0, and we aim to construct an isomorphism $f: S \to S'$ such that $f(0) = 0$.

For a point $v \in \mathbb{S}_L$, consider the sequence $(\lfloor v + kL \rfloor)_{k \geq 0}$. We have seen in the previous section that a similar sequence can be recovered from $\pi(G, v, 0)$. Thus, when trying to construct an isomorphism between G and G', we shall want to preserve this sequence. That is, we will always map a vertex $v \in S$ to a vertex $v' \in S'$ in such a way that $(\lfloor v + kL \rfloor)_{k \geq 0} = (\lfloor v' + kL \rfloor)_{k \geq 0}$. A key observation is that, since $L = \ell/m$ is rational, $(\lfloor v + kL \rfloor)_{k \geq 0}$ is determined by its first m elements. More precisely, the sequence is periodic after subtracting the fixed sequence $(\lfloor kL \rfloor)_{k \geq 0}$. Since there are only finitely many possibilities for the sequence, there are many candidates for such a $v' \in S'$.

To be more precise, let

$$L = \frac{\ell}{m}$$

be irreducible. Then there are precisely ℓ possibilities for the sequence $(\lfloor v + kL \rfloor)_{k \geq 0}$, according to the value of $\lfloor mv \rfloor$. We thus partition $\mathbb{S}_L$ into the arcs

$$[\tfrac{i}{m}, \tfrac{i+1}{m}) \qquad \text{for } 0 \leq i \leq \ell - 1.$$

Note that two points $u \in [\frac{i}{m}, \frac{i+1}{m})$ and $v \in [\frac{j}{m}, \frac{j+1}{m})$ satisfy $(\lfloor u + kL \rfloor)_{k \geq 0} = (\lfloor v + kL \rfloor)_{k \geq 0}$ if and only if $i = j$. We henceforth let $q_v \in \{0, \ldots, \ell - 1\}$ and $r_v \in [0, \frac{1}{m})$ be the unique numbers such that

$$v = \frac{q_v}{m} + r_v.$$

Note that $q_v = \lfloor mv \rfloor$ and $v \in [\frac{q_v}{m}, \frac{q_v+1}{m})$. Moreover, since S is integer-distance-free, $r_v \neq r_u$ for distinct $u, v \in S$, and in particular $r_v \neq 0$ for $v \neq 0$ in S, and similarly for S'.

Let $\bar{S} \subset S$ and $\bar{S}' \subset S'$, and suppose that both contain 0. A bijection $f: \bar{S} \to \bar{S}'$ is called an **extended step-isometry** if $f(0) = 0$ and, for every $u, v \in \bar{S}$, we have

$$q_u = q_{f(u)} \qquad \text{and} \qquad r_u < r_v \iff r_{f(u)} < r_{f(v)}.$$

Though we will not require it, we note that such an extended step-isometry f satisfies that

$$\lfloor m \cdot \|u - v\| \rfloor = \lfloor m \cdot \|f(u) - f(v)\| \rfloor \qquad \text{for all } u, v \in \bar{S}.$$

Proof of Proposition 5.1 Let $S = \{s_n\}_{n\geq 0}$ and $S' = \{s'_n\}_{n\geq 0}$, where $s_0 = s'_0 = 0$. Set $S_0 = S'_0 = \{0\}$ and let $f_0 \colon S_0 \to S'_0$ satisfy $f_0(0) = 0$. For $n \geq 1$, we inductively define a pair of finite sets (S_n, S'_n) and a bijection $f_n : S_n \to S'_n$ such that

- $S_{n-1} \subset S_n \subset S$ and $S'_{n-1} \subset S'_n \subset S'$,
- $s_n \in S_n$ and $s'_n \in S'_n$,
- f_{n-1} is the restriction of f_n to S_{n-1},
- f_n is an extended step-isometry and an isomorphism between $G[S_n]$ and $G'[S'_n]$, where $G[U]$ denotes the subgraph of G induced by U.

The limiting function $\bigcup_n f_n$ will then be the desired isomorphism.

Fix $n \geq 0$ and suppose that we have already constructed (S_n, S'_n) and f_n. To construct (S_{n+1}, S'_{n+1}) and f_{n+1}, we use the back-and-forth method: First, we find a suitable image, say s', for s_{n+1} in S' (this is the 'forth' part). If the image of s_{n+1} was already determined in a previous step, i.e., $s_{n+1} \in S_n$, then we skip this part and set $s' = f_n(s_{n+1})$. Next, we find a suitable preimage, say t, for s'_{n+1} in S (this is the 'back' part). As before, if the preimage of s'_{n+1} was already determined, i.e., $s'_{n+1} \in S'_n \cup \{s'\}$, then we skip this part and set t to be this preimage. We then define $S_{n+1} = S_n \cup \{s_{n+1}, t\}$, $S'_{n+1} = S'_n \cup \{s'_{n+1}, s'\}$ and $f_{n+1} = f_n \cup \{(s_{n+1}, s'), (t, s'_{n+1})\}$. In this way, the first three properties above will automatically hold, so that 'suitable' refers solely to satisfying the fourth property. The two parts are analogous to one another, and so we only explain the first part.

Denote $s := s_{n+1}$ and suppose that $s \notin S_n$. We wish to find a suitable image $s' \in S'$ for s. Thus we need an element of S' such that $f_n \cup \{(s, s')\}$ is an extended step-isometry and an isomorphism between $G[S_n \cup \{s\}]$ and $G'[S'_n \cup \{s'\}]$. Let us first describe those candidates which ensure the former condition.

Consider the values

$$a := \max\{r_{f_n(u)} : u \in S_n \text{ and } r_u < r_s\},$$

$$b := \min\{r_{f_n(u)} : u \in S_n \text{ and } r_u > r_s\} \cup \{\tfrac{1}{m}\}.$$

Note that the set defining a is not empty, since $0 \in S_n$ and $0 = r_0 < r_s$. Since f_n is an extended step-isometry, any element in the set defining a is strictly smaller than any element in the set defining b, and hence $a < b$. Denote

$$I := \frac{q_s}{m} + (a, b).$$

Observe that, since S is integer-distance-free, $I \cap S'$ is precisely the set of $s' \in S'$ such that $f_n \cup \{(s, s')\}$ is an extended step-isometry.

It remains only to show that $I \cap S'$ contains an element s' such that $f_n \cup \{(s, s')\}$ is an isomorphism between $G[S_n \cup \{s\}]$ and $G'[S'_n \cup \{s'\}]$. Since f_n is an isomorphism between $G[S_n]$ and $G'[S'_n]$, it suffices to show that $I \cap S'$ contains an element s' which is adjacent to every element in $f_n(N(s) \cap S_n)$ and to no other element in S'_n. This will follow from the fact that G' is g.e.c. and has unit threshold once we show that $f_n(J)$ is contained in the open arc $(z-1, z+1)$ for some $z \in I \cap S'$, where $J = N(s) \cap S_n$. We will in fact show that this holds for every $z \in I$.

Note that, since G has unit threshold, $J \subset N(s) \subset (s-1, s+1)$. Fix $z \in I$, let $x \in J$ and denote $y := f_n(x)$. We need to show that $y \in (z-1, z+1)$. Recall that $q_y = q_x$ and $q_z = q_s$. Since $x < s+1$, we have $q_x \leq q_s + m$ (if $s > L-1$, so that $q_s + m \geq \ell$, this should be interpreted modulo ℓ).

- If $q_x \neq q_s + m$ then also $q_y < q_z + m$ which implies $y < z+1$.
- If $q_x = q_s + m$, more care is needed. Since $x - s = (q_x - q_s)/m + r_x - r_s < 1$ it follows that $r_x < r_s$. Let $u, v \in S_n$ be the points with $r_{f_n(u)} = a$ and $r_{f_n(v)} = b$. By the definition of a and b, we cannot have $r_x \in (r_u, r_s)$, so $r_x \leq r_u$ and so $r_y \leq a$. Therefore,

$$y = \frac{q_y}{m} + r_y \leq \frac{q_s + m}{m} + a < 1 + z.$$

The other direction (that $y > z-1$) is almost identical. Either $q_x > q_s - m$ and all is well, or else $q_x = q_s - m$ and then $r_x \geq r_v$ and $r_y \geq b$, concluding as above. A small difference is that if $b = 1/m$, there are no points in S_n with $r_x > r_s$, so the latter case is impossible. □

For the most part, the $L = \infty$ case handled in [2] (corresponding to the metric space $(\mathbb{R}, |\cdot|)$) behaves similarly to the case of rational $L > 2$. In particular, both have the Rado property and every dense countable set which is integer-distance-free is Rado. However, for sets which are not integer-distance-free, there are subtle differences between the two cases. As an example, we contrast the set of rationals in $\mathbb{R}$ and in $\mathbb{S}_L$.

Proposition 5.2 *The set of rationals $\mathbb{Q}$ is strongly non-Rado in $(\mathbb{R}, |\cdot|)$.*

This statement—as well as an analogous statement for $(\mathbb{R}^d, \ell_\infty)$—appeared in the proof of Theorem 2(i) in [1]. We give a proof for completeness.

Proof Let $G, G' \in \mathcal{G}_{\infty,\mathbb{Q}}$. Let us first show that any isomorphism $f : \mathbb{Q} \to \mathbb{Q}$ from G to G' must map $x + \mathbb{Z}$ to $f(x) + \mathbb{Z}$. More specifically, there exists $\varepsilon \in \{\pm 1\}$ such that $f(x+n) = f(x) + \varepsilon n$ for all $x \in \mathbb{Q}$ and $n \in \mathbb{Z}$. To see this, observe that x is the unique vertex in G which is at graph-distance 3 in G from both $x-2$ and $x+2$ (by (1), points at distance 3 from x are precisely those in $(x-3, x-2] \cup [x+2, x+3)$). Moreover, if for some u and v there is a unique vertex x with $\text{dist}_G(x, u) = \text{dist}_G(x, v) = 3$, then necessarily $|u - v| = 4$ (otherwise there would be no such vertex or infinitely many). This implies that $\{f(x \pm 2)\} = \{f(x) \pm 2\}$. A similar argument shows that $\{f(x \pm k)\} = \{f(x) \pm k\}$ for any integer $k \geq 2$ and any x. It is easy to deduce from this that for any x there is $\varepsilon_x \in \{\pm 1\}$ such that $f(x+n) = f(x) + \varepsilon_x n$ for all n. For any $x, y \in [0, 1)$, we have by (1) that $\text{dist}_{G'}(f(x+n), f(y+n)) = \text{dist}_G(x+n, y+n) \leq 2$ and $\text{dist}_{G'}(f(x)+n, f(y)-n) \geq |2n| - 1$. Thus, $\varepsilon_x = \varepsilon_y$ for all x and y, which yields our claim.

Now let $G, G' \in \mathcal{G}_{\infty,\mathbb{Q},p}$ be independent and consider two nearby points in G, say 0 and $\frac{1}{2}$. For $u, v \in \mathbb{Q}$, consider the sets

$$A := \{n \in \mathbb{Z} : n \text{ and } \tfrac{1}{2} + n \text{ are adjacent in } G\},$$
$$B^{\pm}_{u,v} := \{n \in \mathbb{Z} : u \pm n \text{ and } v \pm n \text{ are adjacent in } G'\}.$$

By Lemma 2.1 and the above, almost surely, if $G \cong G'$ then $A = B^{+}_{u,v}$ or $A = B^{-}_{u,v}$ for some u and v (namely, for $u = f(0)$ and $v = f(\frac{1}{2})$, where f is an isomorphism from G to G'). However, since $B^{\pm}_{u,v}$ is a sequence of independent Ber(p) random variables, independent also of A, this clearly has probability zero for any fixed u and v. Since there are countably many choices for u and v, we deduce that $\mathbb{P}(G \cong G') = 0$. □

Proposition 5.3 *Let $L > 2$ be rational. Then $\mathbb{Q} \cap \mathbb{S}_L$ is Rado.*

Proof The proof is essentially the same as for integer-distance-free S (Proposition 5.1), with a small twist—instead of finding a suitable image for a single vertex s at a time, we do so for several vertices at a time, those in a certain equivalence class of s.

Let $L = \frac{\ell}{m}$ be irreducible. Say that $u, v \in \mathbb{Q} \cap \mathbb{S}_L$ are equivalent if $\|u - v\|$ is a multiple of $\frac{1}{m}$ and write $[u]$ for the equivalent class of u. Note that $|[u]| = \ell$ and that $[u] = v + \frac{1}{m}\{0, \ldots, \ell - 1\}$ for some $v \in [0, \frac{1}{m})$. We also write $[U] := \bigcup_{u \in U}[u]$.

Let $S = \{s_n\}_{n \ge 0}$ be an enumeration of representatives of $\mathbb{Q} \cap \mathbb{S}_L$, where $s_0 = 0$ and $s_n \in (0, \frac{1}{m})$ for $n \ge 1$. The isomorphism $f : \mathbb{Q} \cap \mathbb{S}_L \to \mathbb{Q} \cap \mathbb{S}_L$ between G and G' that we aim to construct will be defined completely by a permutation of S by requiring that

$$f(s + \tfrac{i}{m}) = f(s) + \tfrac{i}{m} \qquad \text{for all } s \in S \text{ and } 0 \le i \le \ell - 1. \tag{2}$$

We shall also require that $f(0) = 0$. Thus, we only need to prescribe the value of f at one element in each equivalence class, as the rest are then determined by this.

Suppose that we have already constructed a partial permutation of S which gives rise through (2) to a function f which is an extended step-isometry and an isomorphism of the induced subgraphs. That is, f is a bijection between some $S_n \subset S$ and $S'_n \subset S$, and it extends to a bijection from $[S_n]$ to $[S'_n]$ by (2). To proceed with the 'forth' step of the back-and-forth method, we choose some $s \in S \setminus S_n$. We need to find an image $s' \in S \setminus S'_n$ for s such that $f \cup \{(s + \frac{i}{m}, s' + \frac{i}{m}) : 0 \le i < \ell\}$ is an extended step-isometry and an isomorphism between $G[[S_n] \cup [s]]$ and $G'[[S'_n] \cup [s']]$. A similar argument to the one given in the proof of Proposition 5.1 shows that there is an open interval of candidates in $(0, \frac{1}{m})$ which satisfy the extended step-isometry requirement. Since each such candidate satisfies the isomorphism requirement with positive (constant) probability, and since these events are independent, there almost surely exists a suitable choice for s. Other than this difference, the proof proceeds identically. □

Remark 5.4 When $L > 2$ is rational, one can also construct dense countable subsets of $\mathbb{S}_L$ which are neither Rado nor strongly non-Rado, so that the probability to have an isomorphism is neither zero nor one. One such way is to take a dense integer-distance-free set and add the points of $\{\frac{i}{2m} : 0 \le i < 2\ell\}$. We omit the details. See [1, Theorem 2] for a similar statement when the underlying metric space is a finite-dimensional normed space.

Acknowledgement This work was supported in part by NSERC of Canada.

References

1. Balister, P., Bollobás, B., Gunderson, K., Leader, I., Walters, M.: Random geometric graphs and isometries of normed spaces. Trans. Amer. Math. Soc. **370**(10), 7361–7389 (2018). MR 3841851
2. Bonato, A., Janssen, J.: Infinite random geometric graphs. Ann. Comb. **15**(4), 597–617 (2011). MR 2854782
3. Bonato, A., Janssen, J.: Infinite random geometric graphs from the hexagonal metric. In: Combinatorial Algorithms. Lecture Notes in Computer Science, vol. 7643, pp. 6–19. Springer, Heidelberg (2012). MR 3056370
4. Bonato, A., Janssen, J.: Infinite random graphs and properties of metrics. In: Recent Trends in Combinatorics, IMA Vol. Math. Appl., vol. 159, pp. 257–273. Springer, Cham (2016). MR 3526412
5. Bonato, A., Janssen, J., Quas, A.: Geometric random graphs and Rado sets in sequence spaces. Eur. J. Combin. **79**, 1–14 (2019). MR 3895055
6. Bonato, A., Janssen, J., Quas, A.: Geometric random graphs and Rado sets of continuous functions. Discrete Anal. (2021). Paper No. 3, 21. MR 4244349
7. Cameron, P.J.: The random graph. In: The Mathematics of Paul Erdős, II. Algorithms Combin., vol. 14, pp. 333–351. Springer, Berlin (1997). MR 1425227
8. Erdős, P., Rényi, A.: Asymmetric graphs. Acta Math. Acad. Sci. Hungar. **14**, 295–315 (1963). MR 156334
9. Rado, R.: Universal graphs and universal functions. Acta Arith. **9**, 331–340 (1964). MR 172268

More About Asymptotic Properties of Some Binary Classification Methods for High Dimensional Data

Addy Bolivar-Cime

Abstract In this manuscript we study the asymptotic behavior of the following binary classification methods: Support Vector Machine, Mean Difference, Distance Weighted Discrimination and Maximal Data Piling, when the dimension of the data increases and the sample sizes of the classes are fixed. We consider multivariate data with the asymptotic geometric structure of n-simplex, such that the multivariate standard Gaussian distribution, as the dimension increases and the sample size n is fixed. We provide the asymptotic behavior of the four methods in terms of the angle between the normal vector of the separating hyperplane of the method and the optimal direction for classification, under more general conditions than those of Bolivar-Cime and Cordova-Rodriguez (Commun Stat Theory Methods 47(11):2720–2740, 2018). We also analyze the asymptotic behavior of the probabilities of misclassification of the methods. A simulation study is performed to illustrate the theoretical results.

Keywords High dimensional data · Binary discrimination · Asymptotic behavior · Machine learning · Support vector machine

1 Introduction

For random samples from the multivariate standard Gaussian distribution, it is known that as the dimensión d increases while the sample size n is fixed, the data tend to be at vertices of an n-simplex, that is, a regular polyhedron of n vertices, see [5]. This asymptotic geometric representation of the high dimensional standard Gaussian data can be observed for more general distributions, see for example [2, 5, 6, 9, 13].

A. Bolivar-Cime (✉)
División Académica de Ciencias Básicas, Universidad Juárez Autónoma de Tabasco, Villahermosa, Mexico
e-mail: addy.bolivar@ujat.mx

D. Hernández-Hernández et al. (eds.), *Advances in Probability and Mathematical Statistics*, Progress in Probability 79, https://doi.org/10.1007/978-3-030-85325-9_3

Considering an asymptotic geometric structure of high dimensional data, the asymptotic behavior of several binary classification methods have been studied as the dimension increases and the sample sizes of the classes are fixed, some works in this direction are [1, 3–5, 7–10]. These authors studied the asymptotic behavior of binary classification methods in terms of the probability of misclassification or in terms of the angle between the normal vector of the separating hyperplane of the method and the optimal direction for classification, some of the authors consider both approaches.

In the present work we consider the binary classification methods Mean Difference [11], Support Vector Machine [12], Distance Weighted Discrimination [8] and Maximal Data Piling [1]. We provide the asymptotic behavior of the angles between the normal vectors of the separating hyperplanes of the methods and the optimal direction for classification, under more general conditions that those considering in [3], as the dimension tends to infinity and the sample sizes are fixed, when the data of the classes have the asymptotic geometric representation. We also analyze the asymptotic behavior of the probabilities of misclassification of these methods. Finally, we perform a simulation study to illustrate the theoretical results.

The manuscript is divided as follows. In Sect. 2 we provide the asymptotic behavior of the four methods, in terms of the angle between the normal vector of the separating hyperplane and the optimal direction, and in terms of the probability of misclassification. In Sect. 3 we present a simulation study. The conclusions are given in Sect. 4. Finally, in the Appendix we provide some technical details.

2 Asymptotic Behavior of Binary Classification Methods

We consider two independent classes $C_+ = \{X_1, X_2, \ldots, X_m\}$ and $C_- = \{Y_1, Y_2, \ldots, Y_n\}$ of independent and identically distributed d-dimensional random vectors with means μ_+ and μ_- and covariance matrices Σ_+ and Σ_-, respectively, with the asymptotic geometric representation as the dimension increases and the sample sizes are fixed, in the sense that

$$\frac{\| X_i - \mu_+ \|^2}{d} \xrightarrow{P} \sigma^2, \ \forall i; \qquad \frac{\| X_i - X_j \|^2}{d} \xrightarrow{P} 2\sigma^2, \ \forall i \neq j; \tag{1}$$

$$\frac{\| Y_i - \mu_- \|^2}{d} \xrightarrow{P} \tau^2, \ \forall i; \qquad \frac{\| Y_i - Y_j \|^2}{d} \xrightarrow{P} 2\tau^2, \ \forall i \neq j; \tag{2}$$

as $d \to \infty$, for some constants $\sigma > 0$ and $\tau > 0$. Let $v_d = \mu_+ - \mu_-$. Additionally we suppose that

$$\frac{\| v_d \|^2}{d} = \frac{\| \mu_+ - \mu_- \|^2}{d} \longrightarrow c^2, \tag{3}$$

$$\frac{\| X_i - Y_j \|^2}{d} \xrightarrow{P} \sigma^2 + \tau^2 + c^2, \ \forall i, j, \tag{4}$$

as $d \to \infty$, for some constant $c > 0$.

Some conditions that imply (1)–(4) are given in [2, 5, 9]. When (1)–(4) hold, after rescaling by $d^{-1/2}$, the vectors of the classes C_+ and C_- tend to be the vertices of an m-simplex and n-simplex, respectively, as the dimension increases. Since the rescaled distances between the vertices of one simplex and the vertices of the other simplex tend to a constant as the dimension increases, all the data tend to be at the vertices of an N-polyhedron, with $N = n + m$, m of its vertices arranged in an m-simplex and the other n vertices arranged in an n-simplex.

2.1 Angle Between the Normal Vector and the Optimal Direction

Considering the classes C_+ and C_- as before, satisfying (1)–(4), the optimal direction for the normal vector of a separating hyperplane is $v_d = \mu_+ - \mu_-$ as the dimension increases. Therefore, a good linear classification method is expected to have its normal vector of the separating hyperplane approximately in the same direction as the optimal direction. We consider the linear classification methods Mean Difference (MD), Support Vector Machine (SVM), Distance Weighted Discrimination (DWD) and Maximal Data Piling (MDP). In the next theorem we provide the asymptotic behavior of the angles between the normal vectors of the separating hyperplanes of the four methods and the optimal direction for classification, as the dimension tends to infinity and the sample sizes are fixed, when the data of the classes have the asymptotic geometric representation.

Theorem 2.1 *Let the classes C_+ and C_- be as before, satisfying (1)–(4). If v represents the normal vector of the MD, SVM, DWD or MDP hyperplane for the training dataset, then*

$$\text{Angle}(v, v_d) \xrightarrow{P} \arccos\left[\frac{c}{(\sigma^2/m + \tau^2/n + c^2)^{1/2}}\right] \quad \textit{as } d \to \infty.$$

The proof of the last theorem is given in Sect. 4. Observe that if $c \to 0$ the distance, scaled by $d^{-1/2}$, between the class means approaches to zero and, by Theorem 2.1, Angle(v, v_d) approaches to $\pi/2$, that is v and v_d tend to be orthogonal, and we say that the four methods tend to be *strongly inconsistent* in terms of the angle between the normal vector and the optimal direction. On the other hand, if $c \to \infty$ the scaled distance between the class means tends to infinity and $\text{Angle}(v, v_d)$ approaches to 0, that is v and v_d tend to be in the same direction, and in this case we say that the four methods tend to be *consistent* in terms of the angle between the normal vector and the optimal direction. Therefore, the larger the scaled distance between the class means, the smaller the angles between the normal vectors and the optimal direction, when the dimension is large.

In [3] a similar result to Theorem 2.1 is proven. They consider classes C_+ and C_- with the same covariance matrix Σ_d; C_+ and C_- have means v_d and zero, respectively; the data in C_+ and C_- satisfy (1) and (2) with $\sigma = \tau$; v_d satisfies (3); and Σ_d and v_d satisfy $D_{\Sigma_d}(0, v_d)^2/(d \parallel v_d \parallel^2) \to 0$ as $d \to \infty$, where $D_{\Sigma_d}(x, y) = [(x - y)^\top \Sigma_d (x - y)]^{1/2}$ is the Mahalanobis distance corresponding to Σ_d. It is easy to show that the last condition implies (4), therefore the conditions of Theorem 2.1 are more general than those of the result of [3].

Moreover, in [9] a similar result (their Theorem 6) to Theorem 2.1 is proven, but they only consider the DWD method and the weighted Distance-Weighted Discrimination (wDWD), and they take different assumptions. From their results, it can be seen that their assumptions imply the conditions (1)–(4), therefore the conditions of Theorem 2.1 are more general than those of Theorem 6 of [9].

As mentioned before, the conditions (1)–(4), that imply an asymptotic geometric structure of N-polyhedron of the data, can be obtained under several assumptions. Therefore, Theorem 2.1 provides a theoretical explanation of the asymptotic behavior of the four considered methods as the dimension increases while the sample sizes are fixed, in terms of the directions of the normal vectors of the methods, considering a general setting where the data have an asymptotic geometric structure.

The normal vectors of the separating hyperplanes of MD, SVM and DWD are proportional to the difference between two points on the convex hulls of the two classes, see for example [4]. The Lemma 3.1 given in [3] about an explicit asymptotic representation for the normal vectors of the MD, SVM and DWD hyperplanes, still holds under the assumptions of Theorem 2.1, since we only need the asymptotic geometric structure of N-polyhedron of the data, which is given by the conditions (1)–(4). We reformulate that lemma as follows. Let $\alpha = (\alpha_+^\top, \alpha_-^\top)^\top$ be an N-dimensional vector, where $\alpha_+ = (\alpha_{1+}, \alpha_{2+}, \dots, \alpha_{m+})^\top$ and $\alpha_- = (\alpha_{1-}, \alpha_{2-}, \dots, \alpha_{n-})^\top$ are subvectors of α of dimensions m and n, respectively. Let $\mathbf{1}_k$ be the k-dimensional vector of ones.

Lemma 2.1 *Consider the same assumptions as in Theorem 2.1. Let* $\mathbf{X} = [X_1, X_2, \dots, X_m]$ *and* $\mathbf{Y} = [Y_1, Y_2, \dots, Y_n]$. *If* $\widetilde{v} = \mathbf{X}\alpha_+ - \mathbf{Y}\alpha_-$, *with* $\alpha \geq \mathbf{0}$ *and* $\mathbf{1}_m^\top \alpha_+ = \mathbf{1}_n^\top \alpha_- = 1$, *is proportional to the normal vector of the MD, SVM or DWD hyperplane, then*

$$\alpha_{i+} \xrightarrow{P} \frac{1}{m}, \quad \alpha_{i-} \xrightarrow{P} \frac{1}{n}, \tag{5}$$

as $d \to \infty$, *for* $i = 1, 2, \dots, m$ *and* $j = 1, 2, \dots, n$.

The last lemma tell us that, when dimension tends to infinity, the normal vectors of the SVM and DWD hyperplanes tend to be in the same direction as the normal vector of the MD hyperplane. As we will see in the proof of Theorem 2.1, this is also true for the normal vector of the MDP hyperplane.

2.2 *Probability of Misclassification*

From [5] we have, under the same assumptions as in Theorem 2.1, the following results about the probabilities of misclassification of the MD, SVM and DWD methods:

(a) Assume that $\sigma^2/m \geq \tau^2/n$; if need be, interchange X and Y to achieve this. If $c^2 > \sigma^2/m - \tau^2/n$, then the probability that a new datum from either the X-population or the Y-population is correctly classified by the SVM or MD hyperplane converges to 1 as $d \to \infty$. If $c^2 < \sigma^2/m - \tau^2/n$, then with probability converging to 1 as $d \to \infty$ a new datum from either population will be classified by the SVM or MD hyperplane as belonging to the Y-population.
(b) Assume that $\sigma^2/m^{3/2} \geq \tau^2/n^{3/2}$; if need be, interchange X and Y to achieve this. If $c^2 > (n/m)^{1/2}\sigma^2/m - \tau^2/n$, then the probability that a new datum from either the X-population or the Y-population is correctly classified by the DWD hyperplane converges to 1 as $d \to \infty$. If $c^2 < (n/m)^{1/2}\sigma^2/m - \tau^2/n$, then with probability converging to 1 as $d \to \infty$ a new datum from either population will be classified by the DWD hyperplane as belonging to the Y-population.

As we mentioned at the end of Sect. 2.1, under the same assumptions as in Theorem 2.1, the normal vector of the MDP hyperplane tends to be in the same direction as the normal vector of the MD hyperplane as $d \to \infty$. Therefore, taking the intercept of the MDP hyperplane as $b = -v^\top(\overline{X} + \overline{Y})/2$, where v is the normal vector of the MDP hyperplane and $\overline{X}$ and $\overline{Y}$ are the sample means of the X and Y population, respectively, the MDP and MD hyperplanes coincide when $d \to \infty$. Hence, (a) holds for the MDP hyperplane.

Observe that if $m = n$ and $\sigma = \tau$, then for any $c > 0$ the four methods achieve asymptotically correct classification of a new datum from any population as $d \to \infty$. On the other hand, if $m \neq n$ or $\sigma \neq \tau$, then the methods SVM, MD and MDP can have a different asymptotic behavior from the DWD method in terms of the probability of misclassification. For example, let

$$M_1 = \frac{\sigma^2}{m} - \frac{\tau^2}{n} \quad \text{and} \quad M_2 = \left(\frac{n}{m}\right)^{1/2}\frac{\sigma^2}{m} - \frac{\tau^2}{n}, \tag{6}$$

and suppose $n > m$ and $\sigma^2/m > \tau^2/n$, then $M_1 < M_2$ and if $M_1 < c^2 < M_2$ we have by (a) that the SVM, MD and MDP methods achieve asymptotically correct classification of a new datum from any population as $d \to \infty$, while the DWD method achieves asymptotically perfect classification for the Y-population and completely incorrect classification for the X-population. This shows an advantage of the SVM, MD and MDP methods over the DWD method when the dimension increases, since the first three methods classify correctly new data from any population for a wider range of values of c.

As it was noted in [3], we can have the property of inconsistency of a method in terms of the angle between the normal vector of the separating hyperplane and the optimal direction, in the sense that the angle do not tend to zero as the dimension increases, and at the same time we can have the property of consistency of this method in terms of the classification error rate, in the sense that the error rate tends to zero as the dimension tends to infinity. By Theorem 2.1 and the previous results, we have that the four methods are inconsistent in terms of the angle between the normal vector of the hyperplane and the optimal direction for any $c > 0$, and they are consistent in terms of the classification error rate if, for example, $n \geq m$, $\sigma^2/m \geq \tau^2/n$ and $M_1 \leq M_2 < c^2$. However, we observe that as c increases the four methods tend to be consistent in terms of the angle between the normal vector and the optimal direction.

3 Simulation Study

In order to illustrate the theoretical results presented in Sects. 2.1 and 2.2, we consider analogous multivariate data to those considered in [3], but now we take two classes C_+ and C_- satisfying the assumptions given in the beginning of Sect. 2, with $n \neq m$, $\Sigma_+ \neq \Sigma_-$ and $\sigma \neq \tau$.

Consider the d-dimensional random vector $\mathbf{Z} = (\mathbf{Z}^{(1)}, \mathbf{Z}^{(2)}, \ldots, \mathbf{Z}^{(d)})$, with d even, where $\mathbf{Z}^{(i)}$, for $i = 1, 2, \ldots, d/2$ are independent univariate standard Gaussian variables, and where $\mathbf{Z}^{(j)} = \mathbf{Z}^{(i)2} + \mathbf{Z}^{(i)} - 1$, for $j = d/2 + i$ with $i = 1, 2, \ldots, d/2$. It can be seen that $\mathbf{Z}$ has mean zero and covariance matrix

$$\Sigma_d = \begin{bmatrix} I_{d/2} & I_{d/2} \\ I_{d/2} & 3I_{d/2} \end{bmatrix}.$$

As it is shown in [3], if $\mathbf{Z}_1$ and $\mathbf{Z}_2$ are independent random vector with the same distribution as $\mathbf{Z}$, then

$$\frac{\| \mathbf{Z}_i \|^2}{d} \xrightarrow{P} 2 \quad \text{and} \quad \frac{\| \mathbf{Z}_1 - \mathbf{Z}_2 \|^2}{d} \xrightarrow{P} 4, \tag{7}$$

as $d \to \infty$, for $i = 1, 2$.

Let C_+ be the class of independent and identically distributed random vectors $X_1, X_2, \ldots, X_m$, with the same distribution as $r\mathbf{Z} + \beta \mathbf{1}_d$, where $r > 0$ and $\beta > 0$. Let C_- be the class of independent and identically distributed random vectors $Y_1, Y_2, \ldots, Y_n$, with the same distribution as $\mathbf{Z}$. Then the classes C_+ and C_- have means $\mu_+ = \beta \mathbf{1}_d$ and $\mu_- = \mathbf{0}$, and covariance matrices $\Sigma_+ = r^2 \Sigma_d$ and $\Sigma_- = \Sigma_d$, respectively. Suppose that C_+ and C_- are independent. In the Appendix, it is shown that C_+ and C_- satisfy the conditions (1)–(4), with $\sigma = \sqrt{2}r$, $\tau = \sqrt{2}$ and $c = \beta$.

We considered the classes C_+ and C_- with $m = 16$, $n = 40$, $r = 2$ and the dimensions $d = 100, 500, 1500, 2500$. The thresholds of c^2 for the consistency, in terms of the classification error rate, of the methods SVM, MD, MDP and DWD, given by (6), are equal to $M_1 = 9/20 = 0.45$ and $M_2 = \sqrt{2.5}/2 - 1/20 \approx 0.74$. Since $M_1^{1/2} \approx 0.67$ and $M_2^{1/2} \approx 0.86$, we took $c = 0.5, 0.75$ and 1 in order to have $c < M_1^{1/2} < M_2^{1/2}$, $M_1^{1/2} < c < M_2^{1/2}$ and $M_1^{1/2} < M_2^{1/2} < c$, respectively.

For each value of d and c, we generated 500 training datasets (of sizes $m = 16$ and $n = 40$ for the classes C_+ and C_-, respectively), then we calculated the means of the angles between the normal vectors of the four methods and the optimal direction $v_d = \mu_+ - \mu_- = \beta \mathbf{1}_d$. We also calculated the means of the classification error rates of the four methods taking test dataset of size 100 for each class.

We summarize the results of the simulations in the Figs. 1, 2, and 3, and we observe the following:

- For the case $c = 0.5$, by Theorem 2.1 the angles between the normal vectors of the four methods and the optimal direction converge in probability to 0.9776 as $d \to \infty$, and we observe in (a) of Fig. 1 that as the dimension increases the means of angles approach to that value. On the other hand, by the results mentioned in Sect. 2.2, since $c < M_1^{1/2} < M_2^{1/2}$, the four methods achieve asymptotically perfect classification for the Y-population and completely incorrect classification for the X-population as $d \to \infty$, and this is observed in (b) of Fig. 1 since for the four methods the means of the classification error rates approach to 0.5 as the dimension increases. In this case the four methods are inconsistent in terms of the angle between the normal vector and the optimal direction, and they are also inconsistent in terms or the classification error rate.
- For $c = 0.75$, by Theorem 2.1 the angles between the normal vectors of the four methods and the optimal direction converge in probability to 0.7798 as $d \to$

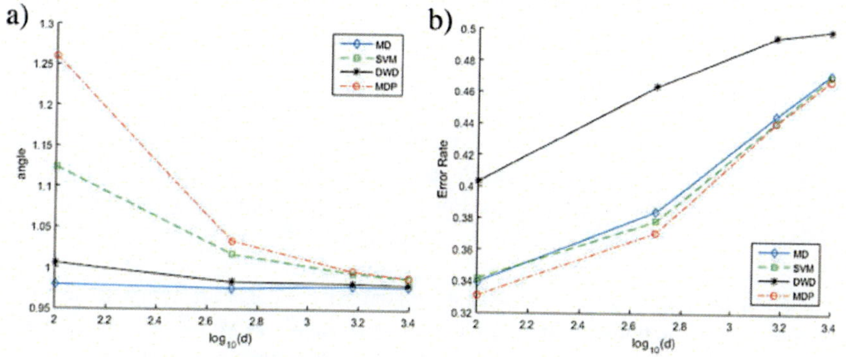

Fig. 1 (**a**) Means of the angles between the normal vectors and the optimal direction, (**b**) means of the classification error rates, for the case $c = 0.5$

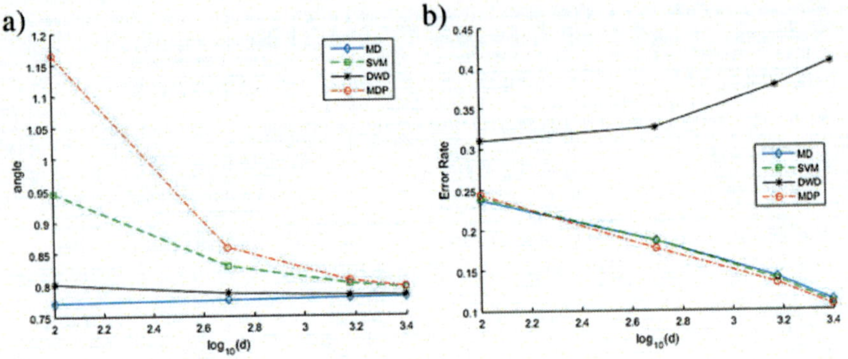

Fig. 2 (**a**) Means of the angles between the normal vectors and the optimal direction, (**b**) means of the classification error rates, for the case $c = 0.75$

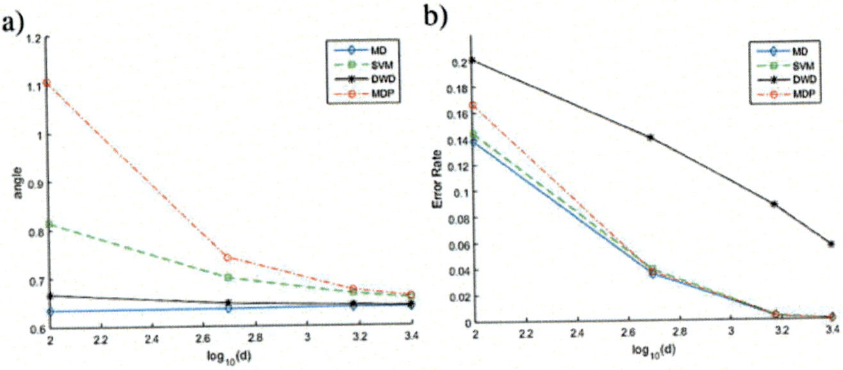

Fig. 3 (**a**) Means of the angles between the normal vectors and the optimal direction, (**b**) means of the classification error rates, for the case $c = 1$

∞, and this agrees with (a) of Fig. 2, where we observe that as the dimension increases the means of the angles approach to that value. By the results mentioned in Sect. 2.2, since in this case $M_1^{1/2} < c < M_2^{1/2}$, the SVM, MD and MDP methods achieve asymptotically correct classification of a new datum from any population as $d \to \infty$, and DWD achieves asymptotically perfect classification for the Y-population and completely incorrect classification for the X-population. This is observed in (b) of Fig. 2, where the means of the error rates of the SVM, MD and MDP methods decrease toward zero as the dimension increases, while the mean of the error rate of DWD increases toward 0.5. Therefore, in this case the four methods are inconsistent in terms of the angle between the normal vector and the optimal direction. On the other hand, SVM, MD and MDP are consistent while DWD is inconsistent in terms or the classification error rate.

- For the case $c = 1$, by Theorem 2.1 the angles between the normal vectors of the four methods and the optimal direction converge in probability to 0.6381 as $d \to \infty$, which is observed in (a) of Fig. 3, since the means of the angles approach to that value as the dimension increases. By the results in Sect. 2.2, since in this case $M_1^{1/2} < M_2^{1/2} < c$, the four methods achieve asymptotically correct classification of a new datum from any population as $d \to \infty$, and this is observed in (b) of Fig. 3 where the error rates approach zero as the dimension increases. Hence, the four methods are inconsistent in terms of the angle between the normal vector and the optimal direction, however they are consistent in terms of the classification error rate.
- With respect to the angles between the normal vectors of the methods and the optimal direction, we observe that for all the considered values of d and c, the MD method has the smaller angles, followed by DWD, SVM and MDP.
- For the considered values of c, we observe that the SVM, MD and MDP methods have very similar error rates, and this is because their hyperplanes asymptotically coincide as the dimension tends to infinity. Furthermore, the error rates of these three methods are smaller than the error rate of the DWD method.

4 Conclusions

In this manuscript we showed that when data of the two classes C_+ and C_- have the asymptotic geometric structure of N-polyhedron, with $N = m + n$, where m of the vertices correspond to an m-simplex and the remaining n vertices correspond to an n-simplex, the angles between the normal vectors of the separating hyperplanes of the MD, SVM, DWD and MDP methods and the optimal direction for classification, converge in probability to the same constant as the dimension d tends to infinity. The limit in probability of the angles depends on the limit of the distance, scaled by $d^{-1/2}$, between the class means. As the limit of this scaled distance increases toward infinity, the limit in probability of the angles approaches to zero, therefore we say that the four methods tend to be asymptotically consistent in terms of the angle between the normal vector and the optimal direction; as the limit of the scaled distance between the class means decreases toward zero, the limit in probability of the angles approaches to $\pi/2$, therefore we say that the four methods tend to be asymptotically strongly inconsistent in terms of the angle of the normal vector and the optimal direction. This result is similar to the one given in [3], but now we consider more general assumptions.

It is interesting that these four methods can be inconsistent in terms of the angle between the normal vector and the optimal direction, in the sense that this angle do not tend to zero as the dimension increases, and at the same time the four methods can be consistent in terms of the classification error rate, in the sense that this error rate tends to zero as the dimension increases. This phenomenon is due to the asymptotic geometric structure of the data.

We observed that even when the four methods have the same asymptotic behavior in terms of the angle between the normal vector and the optimal direction, and the normal vectors of the four methods tend to be in the same direction, the SVM, MD and MDP methods can have a very different behavior from the DWD method in terms of the classification error rate. This is because the separating hyperplanes of SVM, MD and MDP asymptotically coincide as the dimension increases, while the separating hyperplane of DWD can be asymptotically different from the others under some conditions.

Appendix

Proof of Theorem 2.1

Case 1: The vector v is the normal vector of the MD, SVM or DWD hyperplane. Let $\widetilde{v}$ be the vector given in Lemma 2.1. Let $X_i' = X_i - \mu_+$ and $Y_j' = Y_j - \mu_-$, for $i = 1, 2, \ldots, m$ and $j = 1, 2, \ldots, n$. We denote by $\langle x, y\rangle$ the dot product between the d-dimensional vectors x and y. Note that

$$\begin{aligned}\| \widetilde{v} \|^2 &= \sum_{i=1}^{m} \alpha_{i+}^2 \| X_i' \|^2 + \sum_{j=1}^{n} \alpha_{j-}^2 \| Y_j' \|^2 + \| v_d \|^2 \\ &+ 2\sum_{i<j} \alpha_{i+}\alpha_{j+}\left\langle X_i', X_j'\right\rangle + 2\sum_{i<j} \alpha_{i-}\alpha_{j-}\left\langle Y_i', Y_j'\right\rangle \\ &+ 2\sum_{i=1}^{m} \alpha_{i+}\left\langle X_i', v_d\right\rangle - 2\sum_{j=1}^{n} \alpha_{j-}\left\langle Y_j', v_d\right\rangle - 2\sum_{i=1}^{m}\sum_{j=1}^{n} \alpha_{i+}\alpha_{j-}\left\langle X_i', Y_j'\right\rangle. \end{aligned} \tag{8}$$

Dividing both sides of (8) by d, we have by Lemma 2.1 and (1)–(3) that the sum of the first three terms of the right side converges in probability to $\sigma^2/m+\tau^2/n+c^2$ as $d \to \infty$. Now we will see that the sum of the last five terms converges in probability to zero as $d \to \infty$. By (1) we have that

$$\frac{\left\langle X_i', X_j'\right\rangle}{d} = \frac{1}{2}\left(\frac{\| X_i' \|^2}{d} + \frac{\| X_j' \|^2}{d} - \frac{\| X_i - X_j \|^2}{d}\right) \xrightarrow{P} \frac{1}{2}(\sigma^2 + \sigma^2 - 2\sigma^2) = 0 \tag{9}$$

as $d \to \infty$, for $i \neq j$. Analogously

$$\frac{\left\langle Y_i', Y_j'\right\rangle}{d} \xrightarrow{P} 0 \quad \text{as } d \to \infty, \quad \text{for } i \neq j. \tag{10}$$

Observe that sum of the last three terms of the right side of (8) is equal to

$$2\sum_{i=1}^{m}\sum_{j=1}^{n}\alpha_{i+}\alpha_{j-}[\langle X_i', v_d\rangle - \langle Y_j', v_d\rangle - \langle X_i', Y_j'\rangle].$$

From (1)–(4) and the equality

$$\frac{\| X_i - Y_i \|^2}{d} = \frac{\| X_i' \|^2}{d} + \frac{\| Y_j' \|^2}{d} + \frac{\| v_d \|^2}{d} + \frac{2}{d}\left[\langle X_i', v_d\rangle - \langle Y_j', v_d\rangle - \langle X_i', Y_j'\rangle\right],$$

we have that

$$\frac{1}{d}[\langle X_i', v_d\rangle - \langle Y_j', v_d\rangle - \langle X_i', Y_j'\rangle] \xrightarrow{P} 0 \quad \text{as } d \to \infty,\ \forall i, j. \tag{11}$$

Therefore, by Lemma 2.1 and (9)–(11), we have that the sum of the last five terms of the right side of (8) divided by d converges in probability to zero as $d \to \infty$. Thus,

$$\frac{\| \widetilde{v} \|^2}{d} \xrightarrow{P} \frac{\sigma^2}{m} + \frac{\tau^2}{n} + c^2 \quad \text{as } d \to \infty. \tag{12}$$

From the results of [5], under the asymptotic geometric structure of the data, if $Y_1^*, Y_2^*, \ldots, Y_k^*$ are independent and identically distributed d-dimensional random vectors with the same distribution as the vectors of the class C_- and $\overline{Y}_k^* = \sum_{j=1}^{k} Y_j^*/k$, we have

$$\frac{\| X_i - \overline{Y}_k^* \|^2}{d} \xrightarrow{P} c^2 + \sigma^2 + \frac{\tau^2}{k} \quad \text{as } d \to \infty.$$

Since $\overline{Y}_k^*$ converges in probability to μ_- as $k \to \infty$, we have that

$$\frac{\| X_i - \mu_- \|^2}{d} \xrightarrow{P} c^2 + \sigma^2 \quad \text{as } d \to \infty.$$

Furthermore, by (1) and (3)

$$\frac{\| X_i - \mu_+ \|^2}{d} \xrightarrow{P} \sigma^2 \quad \text{and} \quad \frac{\| \mu_+ - \mu_- \|^2}{d} \xrightarrow{P} c^2,$$

as $d \to \infty$. Thus by the Pythagoras theorem, after rescaling by $d^{-1/2}$, the segments $X_i\mu_-$, $X_i\mu_+$ and $\mu_+\mu_-$ tend to form a right triangle as $d \to \infty$, where the hypotenuse is $X_i\mu_-$. Therefore, $X_i'/d^{1/2} = (X_i - \mu_+)/d^{1/2}$ and $v_d/d^{1/2} = (\mu_+ - \mu_-)/d^{1/2}$ tend to be orthogonal as $d \to \infty$, then

$$\frac{\langle X_i', v_d\rangle}{d} \xrightarrow{P} 0 \quad \text{as } d \to \infty,\ \forall i. \tag{13}$$

We also have that

$$\frac{\langle X_i', v_d\rangle}{d^{1/2} \parallel v_d \parallel} = \frac{\langle X_i', v_d\rangle}{d} \frac{d^{1/2}}{\parallel v_d \parallel} \xrightarrow{P} 0 * c^{-1} = 0 \quad \text{as } d \to \infty,\ \forall i. \tag{14}$$

Analogously,

$$\frac{\langle Y_j', v_d\rangle}{d} \xrightarrow{P} 0, \quad \frac{\langle Y_j', v_d\rangle}{d^{1/2} \parallel v_d \parallel} \xrightarrow{P} 0, \quad \text{as } d \to \infty,\ \forall j. \tag{15}$$

Note that

$$\langle \widetilde{v}, v_d\rangle = \sum_{i=1}^{m} \alpha_{i+} \langle X_i', v_d\rangle - \sum_{j=1}^{n} \alpha_{j-} \langle Y_j', v_d\rangle + \parallel v_d \parallel^2 . \tag{16}$$

Therefore, dividing both sides of (16) by $d^{1/2} \parallel v_d \parallel$, from Lemma 2.1, (3), (14) and (15) we have

$$\frac{\langle \widetilde{v}, v_d\rangle}{d^{1/2} \parallel v_d \parallel} \xrightarrow{P} c \quad \text{as } d \to \infty. \tag{17}$$

By (12) and (17) we have

$$\frac{\langle \widetilde{v}, v_d\rangle}{\parallel \widetilde{v} \parallel \parallel v_d \parallel} = \frac{\langle \widetilde{v}, v_d\rangle /(d^{1/2} \parallel v_d \parallel)}{\parallel \widetilde{v} \parallel /d^{1/2}} \xrightarrow{P} \frac{c}{(\sigma^2/m + \tau^2/n + c^2)}$$

as $d \to \infty$. Then

$$\text{Angle}(\widetilde{v}, v_d) = \arccos\left(\frac{\langle \widetilde{v}, v_d\rangle}{\parallel \widetilde{v} \parallel \parallel v_d \parallel}\right) \xrightarrow{P} \arccos\left[\frac{c}{(\sigma^2/m + \tau^2/n + c^2)^{1/2}}\right]$$

as $d \to \infty$.

Case 2: The vector v is the normal vector of the MDP hyperplane. Let X_i' and Y_j' be as in case 1, for $i = 1, \ldots, m$ and $j = 1, 2, \ldots, n$. From (1) and the results of [5] we have

$$\frac{\| \overline{X} - \mu_+ \|^2}{d} \xrightarrow{P} \frac{\sigma^2}{m}, \tag{18}$$

$$\frac{\| X_i - \overline{X} \|^2}{d} \xrightarrow{P} \frac{m-1}{m}\sigma^2, \tag{19}$$

as $d \to \infty$. By (11), (13) and (15) we have

$$\frac{\left\langle X_i', Y_j' \right\rangle}{d} \xrightarrow{P} 0 \quad \text{as } d \to \infty, \ \forall i, j. \tag{20}$$

Note that

$$\begin{aligned}\langle X_i - \overline{X}, \overline{X} - \overline{Y} \rangle &= \langle X_i' - (\overline{X} - \mu_+), (\overline{X} - \mu_+) - (\overline{Y} - \mu_-) + v_d \rangle \\ &= \frac{1}{m} \| X_i' \|^2 + \frac{1}{m} \sum_{j \neq i} \left\langle X_i', X_j' \right\rangle - \langle X_i', \overline{Y} - \mu_- \rangle + \langle X_i', v_d \rangle \\ &\quad - \| \overline{X} - \mu_+ \|^2 + \langle \overline{X} - \mu_+, \overline{Y} - \mu_- \rangle - \langle \overline{X} - \mu_+, v_d \rangle .\end{aligned}$$

Dividing both sides of the last equality by $d^{1/2} \| v_d \|$, from (1), (9), (13), (18) and (20) we have

$$\frac{\langle X_i - \overline{X}, \overline{X} - \overline{Y} \rangle}{d^{1/2} \| v_d \|} = \frac{d^{1/2}}{\| v_d \|} \frac{\langle X_i - \overline{X}, \overline{X} - \overline{Y} \rangle}{d} \xrightarrow{P} c^{-1} \left(\frac{\sigma^2}{m} - \frac{\sigma^2}{m} \right) = 0 \tag{21}$$

as $d \to \infty$. Furthermore, by (3) and (12) we have

$$\frac{\| \overline{X} - \overline{Y} \|}{\| v_d \|} \xrightarrow{P} \frac{(\sigma^2/m + \tau^2/n + c^2)^{1/2}}{c} \quad \text{as } d \to \infty. \tag{22}$$

Thus, by (19), (21) and (22) it follows that

$$\begin{aligned}\text{Angle}(X_i - \overline{X}, \overline{X} - \overline{Y}) &= \arccos \left[\frac{\langle X_i - \overline{X}, \overline{X} - \overline{Y} \rangle}{\| X_i - \overline{X} \| \| \overline{X} - \overline{Y} \|} \right] \\ &= \arccos \left[\frac{\langle X_i - \overline{X}, \overline{X} - \overline{Y} \rangle / (d^{1/2} \| v_d \|)}{(\| X_i - \overline{X} \| / d^{1/2})(\| \overline{X} - \overline{Y} \| / \| v_d \|)} \right] \\ &\xrightarrow{P} \arccos(0) = \frac{\pi}{2}\end{aligned} \tag{23}$$

as $d \to \infty$, $\forall i$. Analogously

$$\text{Angle}(Y_j - \overline{Y}, \overline{X} - \overline{Y}) \xrightarrow{P} \frac{\pi}{2} \quad \text{as } d \to \infty,\ \forall j. \tag{24}$$

As it was shown in the proof of Theorem 3.1 of [3], (23) and (24) imply that when d is large, the normal vector v of the MDP hyperplane is approximately in the same direction as $(\overline{X} - \overline{Y})/\parallel \overline{X} - \overline{Y} \parallel$. Hence, $\text{Angle}(v, v_d) = \arccos(\langle v, v_d\rangle/(\parallel v \parallel \parallel v_d \parallel))$ is approximately

$$\arccos[\langle \overline{X} - \overline{Y}, v_d\rangle/(\parallel \overline{X} - \overline{Y} \parallel \parallel v_d \parallel)] \tag{25}$$

when d is large, which by case 1 converges in probability to $\arccos\left[\frac{c}{(\sigma^2/m+\tau^2/n+c^2)^{1/2}}\right]$ as $d \to \infty$.

The Data in the Simulations Satisfy Conditions (1)–(4)

We have that X_i is equal in distribution to $rZ_i + \beta\mathbf{1}_d$, for $i = 1, 2, \ldots, m$, and Y_j is equal in distribution to Z_{m+j}, for $j = 1, 2, \ldots, n$, where $Z_1, Z_2, \ldots, Z_{m+n}$ are independent and identically distributed with the same distribution as the random vector Z given at the beginning of Sect. 3. Therefore, by (7)

$$\frac{\parallel X_i - \mu_+ \parallel^2}{d} = \frac{\parallel Z_i \parallel^2 r^2}{d} \xrightarrow{P} 2r^2, \ \forall i;$$

$$\frac{\parallel X_i - X_j \parallel^2}{d} = \frac{\parallel Z_i - Z_j \parallel^2 r^2}{d} \xrightarrow{P} 4r^2, \ \forall i \neq j;$$

$$\frac{\parallel Y_i - \mu_- \parallel^2}{d} = \frac{\parallel Z_{m+i} \parallel^2}{d} \xrightarrow{P} 2, \ \forall i;$$

$$\frac{\parallel Y_i - Y_j \parallel^2}{d} = \frac{\parallel Z_{m+i} - Z_{m+j} \parallel^2}{d} \xrightarrow{P} 4, \ \forall i \neq j;$$

as $d \to \infty$. Thus conditions (1) and (2) hold with $\sigma = \sqrt{2}r$ and $\tau = \sqrt{2}$. We also have

$$\frac{\parallel v_d \parallel^2}{d} = \frac{\parallel \mu_+ - \mu_- \parallel^2}{d} = \frac{\parallel \beta\mathbf{1}_d \parallel^2}{d} = \beta^2, \tag{26}$$

therefore condition (3) holds with $c = \beta$.

Now we will see that condition (4) holds. Observe that

$$\frac{\| X_i - Y_j \|^2}{d} = \frac{\| Z_i \|^2}{d} r^2 + \frac{\| Z_{m+j}^2 \|}{d} + \frac{\| v_d \|^2}{d}$$
$$- 2r\frac{\langle Z_i, Z_{m+j} \rangle}{d} + 2r\frac{\langle Z_i, v_d \rangle}{d} - 2\frac{\langle Z_{m+j}, v_d \rangle}{d}, \tag{27}$$

for all i, j. From the properties of Z given in [3], we have that

$$\frac{\langle Z_i, Z_j \rangle}{d} \xrightarrow{P} 0, \ \forall i \neq j, \qquad \frac{\langle Z_i, v_d \rangle}{d} \xrightarrow{P} 0, \ \forall i, \quad \text{as } d \to \infty. \tag{28}$$

Therefore, by (7) and (26)–(28) we have

$$\frac{\| X_i - Y_j \|^2}{d} \xrightarrow{P} 2r^2 + 2 + \beta^2 = \sigma^2 + \tau^2 + c^2 \quad \text{as } d \to \infty.$$

Then the condition (4) holds.

References

1. Ahn, J., Marron, J.S.: The maximal data piling direction for discrimination. Biometrika **97**(1), 254–259 (2010)
2. Ahn, J., Marron, J.S., Muller, K.M., Chi, Y.: The high-dimension, low-sample-size geometric representation holds under mild conditions. Biometrika **94**(3), 760–766 (2007)
3. Bolivar-Cime, A., Cordova-Rodriguez, L.M.: Binary discrimination methods for high dimensional data with a geometric representation. Commun. Stat. Theory Methods **47**(11), 2720–2740 (2018)
4. Bolivar-Cime, A., Marron, J.S.: Comparison of binary discrimination methods for high dimension low sample size data. J. Multivar. Anal. **115**, 108–121 (2013)
5. Hall, P., Marron, J.S., Neeman, A.: Geometric representation of high dimension, low sample size data. J. R. Stat. Soc. B **67**(3), 427–444 (2005)
6. Jung, S., Marron, J.S.: PCA consistency in high dimension, low sample size context. Ann. Stat. **37**(6B), 4104–4130 (2009)
7. Marron, J.S.: Distance-weighted discrimination. WIREs Comput. Stat. **7**, 109–114 (2015)
8. Marron, J.S., Todd, M.J., Ahn, J.: Distance-weighted discrimination. J. Am. Stat. Assess. **102**(480), 1267–1271 (2007)
9. Qiao, X., Zhang, H.H., Liu, Y., Todd, M.J., Marron, J.S.: Weighted distance weighted discrimination and its asymptotic properties. J. Am. Stat. Assess. **105**(489), 401–414 (2010)
10. Qiao, X., Zhang, L.: Flexible high-dimensional classification machines and their asymptotic properties. J. Mach. Learn. Res. **16**, 1547–1572 (2015)
11. Scholkopf, B., Smola, A.J.: Learning with Kernels: Support Vector Machines, Regularization, Optimization and Beyond. The MIT Press, Cambridge, Massachusetts (2002)
12. Vapnik, V.N.: The Nature of Statistical Learning Theory. Springer, Berlin (1995)
13. Yata, K., Aoshima, M.: Effective PCA for high-dimension, low-sample-size data with noise reduction via geometric representations. J. Multivar. Anal. **105**(1), 193–215 (2012)

Transport Distances on Random Vectors of Measures: Recent Advances in Bayesian Nonparametrics

Marta Catalano, Antonio Lijoi, and Igor Prünster

Abstract Random vectors of measures are at the core of many recent developments in Bayesian nonparametrics. For a deep understanding of these infinite-dimensional discrete random structures and their impact on the inferential and theoretical properties of the induced models, we consider a class of transport distances based on the Wasserstein distance. The geometrical definition makes it ideal for measuring similarity between distributions with possibly different supports. Moreover, when applied to random vectors of measures with independent increments (*completely random vectors*), the interesting theoretical properties are coupled with analytical tractability. This leads to a new measure of dependence for completely random vectors and the quantification of the impact of hyperparameters in notable models for exchangeable time-to-event data.

Keywords Bayesian nonparametrics · Completely random measures · Completely random vectors · Compound Poisson approximation · Dependence · Lévy copula · Partial exchangeability · Wasserstein distance

1 Introduction

Many notable Bayesian nonparametric models allow to make inference for partially exchangeable sequences. Thanks to de Finetti's representation theorem, the law of any such sequence may be specified in terms of a random vector of probabilities

M. Catalano (✉)
ESOMAS Department, University of Torino, Italy
e-mail: marta.catalano@unito.it

A. Lijoi · I. Prünster
Department of Decision Sciences, Bocconi Institute for Data Science and Analytics (BIDSA), Bocconi University, Milano, Italy
e-mail: antonio.lijoi@unibocconi.it; igor.pruenster@unibocconi.it

D. Hernández-Hernández et al. (eds.), *Advances in Probability and Mathematical Statistics*, Progress in Probability 79, https://doi.org/10.1007/978-3-030-85325-9_4

$(\tilde{P}_1, \ldots, \tilde{P}_m)$. Let $(X_{i,j})_{j\geq 1}$, with $i = 1, \ldots, m$, be a partially exchangeable sequence on $\mathbb{X}$. Then,

$$(X_{i_1,j_1}, \ldots, X_{i_k,j_k})|(\tilde{P}_1, \ldots, \tilde{P}_m) \sim \tilde{P}_{i_1} \times \cdots \times \tilde{P}_{i_k}; \qquad (\tilde{P}_1, \ldots, \tilde{P}_m) \sim Q;$$

for any $k \geq 1$, $i_\ell \in \{1, \ldots, m\}$, $j_\ell \in \mathbb{N} \setminus \{0\}$ such that $(i_\ell, j_\ell) \neq (i_{\ell'}, j_{\ell'})$, for $\ell \neq \ell' = 1, \ldots, k$. When $m = 1$ or $\tilde{P}_1 = \cdots = \tilde{P}_m$ almost surely (a.s.), the model degenerates to exchangeability, which can thus be seen as a special case. There have been many proposals on how to specify the law Q by modeling the dependence structure between random probabilities [21]. Among the most successful specifications, many build on random vectors of measures with independent increments $(\tilde{\mu}_1, \ldots, \tilde{\mu}_m)$, which we denote as *completely random vectors* (CRVs) in analogy with the one-dimensional case of completely random measures (CRMs). Completely random vectors have appealing properties in terms of analytical tractability, typically because of the existence of a multivariate Lévy intensity that characterizes their distribution. For this reason, many random vectors of probabilities may be derived from suitable transformation of CRVs, including normalization [22], kernel mixtures for densities [9, 18] and hazards [7] and exponential transformation for survival functions [6].

The derived nonparametric models for partially exchangeable sequences [8, 10, 15–17, 23] are very flexible but often difficult to interpret, making the prior elicitation more demanding. In order to ease the interpretation and foremost the comparison between different models, we introduce a distance between CRVs, based on the Wasserstein distance. The relationship between the Wasserstein distance and optimal transport theory [25] sheds light on its intrinsically geometric definition. This makes the Wasserstein distance an ideal measure of discrepancy between distributions with possibly different support, in contrast to other common choices, such as the total variation distance, the Hellinger distance and the Kullback–Leibler divergence.

To date the transport distance between CRVs has been used in two different scenarios: to create a new measure of dependence for partially exchangeable models [3] and to measure the discrepancy between hazard rates models for exchangeable observations [2]. The dependence between random measures regulates the borrowing of information between different groups of observations with a major impact on the posterior inference. In order to elicit the prior, one needs a measure of dependence that can be expressed in terms of the hyperparameters of the model. State-of-the-art measures typically consist in linear correlation, thus capturing only a portion of the dependence structure. By leveraging the transport distance between CRVs, Catalano et al. [3] propose a new measure of dependence that goes beyond linear correlation. On the other hand, the transport distance between CRMs, i.e. one-dimensional CRVs, has been fruitfully used in the context of survival analysis. One of the most popular Bayesian nonparametric models for time-to-event data [7] represents the hazard rate function as a kernel mixture over a CRM. For a careful prior elicitation, Catalano et al. [2] find the analytical expression for the Wasserstein distance between the hazards, as the hyperparameters of the CRMs and the kernels vary. When treating the kernel of [7], i.e. $k(t|x) = \beta(x)\mathbb{1}_{(0,t]}$, calculations are performed only for constant $\beta(x) = \beta > 0$, which is often used in applications. This assumption implies that the index of dispersion of the induced hazard $\tilde{h}(t)$ is constant in time, which is often too restrictive. We thus consider the case where $\beta(x)$ increases linearly and compare it to the constant scenario. We find informative bounds on the Wasserstein distance between these two specifications that show how the distance increases quadratically in time.

The work is structured as follows. In Sect. 2 we introduce completely random vectors and in Sect. 3 we define a class of transport distances on them. In Sect. 4 we describe how these distances may be used to define a measure of dependence, reviewing the recent results of [3], whereas in Sect. 5 we focus on its applications in survival analysis, following [2]. New results on time-varying kernels are contained in Sect. 6.

2 Completely Random Vectors

In this section we recall the definition of completely random vectors and their main properties. Let $\mathbb{X}$ be a Polish space with Borel σ-algebra $\mathcal{X}$. We denote by $M_{\mathbb{X}}$ the space of boundedly finite measures on $\mathbb{X}$, endowed with the weak$^\sharp$ topology [5] and the corresponding Borel σ-algebra. An m-dimensional random vector of measures is a measurable function $\tilde{\mu} : \Omega \to M_{\mathbb{X}}^m$, where $(\Omega, \Sigma, \mathbb{P})$ is some probability space and $M_{\mathbb{X}}^m = \prod_{i=1}^m M_{\mathbb{X}}$ denotes the m-fold product space with corresponding product topology and induced Borel σ-algebra. Let $\pi_i : M_{\mathbb{X}}^m \to M_{\mathbb{X}}$ be the i-th projection, i.e. $\pi_i(\mu_1, \dots, \mu_m) = \mu_i$, for $i = 1, \dots, m$. We denote the marginal random measures $\tilde{\mu}_i = \pi_i \circ \tilde{\mu} : \Omega \to M_{\mathbb{X}}$, so that $\tilde{\mu} = (\tilde{\mu}_1, \dots, \tilde{\mu}_m)$.

Definition 2.1 A random vector of measures $\tilde{\mu}$ is said to be a *completely random vector* (CRV) if for every disjoint collection of bounded Borel sets $A_1, \dots, A_n$, the one-dimensional distributions $\tilde{\mu}(A_1), \dots, \tilde{\mu}(A_n)$ are independent.

We observe that for $i = 1, \dots, m$, the marginal random measure $\tilde{\mu}_i$ of a CRV $\tilde{\mu}$ is a completely random measure (CRM) in the sense of [14]. Thus, we can look at CRVs $\tilde{\mu} = (\tilde{\mu}_1, \dots, \tilde{\mu}_m)$ as vectors of dependent CRMs. This property makes them particularly appealing, since dependent CRMs offer the ground for most tractable nonparametric priors in presence of multiple populations.

We focus on CRVs without fixed atoms. Thanks to [13, Theorem 3.19], this ensures the existence of a Poisson random measure $\mathcal{N}$ on $\mathbb{R}_+^m \times \mathbb{X}$ s.t. for every $A \in \mathcal{X}$,

$$\tilde{\mu}(A) \stackrel{\mathrm{d}}{=} \int_{\mathbb{R}_+^m \times A} s\, \mathcal{N}(ds_1, \dots, ds_m, dx), \tag{2.1}$$

where $\stackrel{d}{=}$ denotes equality in distribution and $s = (s_1, \dots, s_m)$. It follows that the distribution of a CRV $\tilde{\mu}$ is characterized by a multivariate Lévy intensity $\nu(ds_1, \dots, ds_m, dx) = \mathbb{E}(\mathcal{N}(ds_1, \dots, ds_m, dx))$ such that (1) $\nu(\mathbb{R}_+^m \times \{x\}) = 0$ for every $x \in \mathbb{X}$; (2) for every bounded $A \in \mathcal{X}$ and every $\epsilon > 0$,

$$\int_{\mathbb{R}_+^m \times A} \min\{s_1 + \cdots + s_m, \epsilon\}\, \nu(ds_1, \dots, ds_m, dx) < +\infty. \tag{2.2}$$

We will focus on infinitely active CRVs, i.e. such that for every Borel set A, the Lévy measures of the marginal CRMs satisfy

$$\int_{\mathbb{R}_+ \times A} \nu_i(ds_i, dx) = \int_{\mathbb{R}_+^m \times A} \nu(ds_1, \dots, ds_m, dx) = +\infty. \tag{2.3}$$

In the next section we define a class of distances between laws of CRVs, whose analytical tractability heavily relies on the existence of multivariate Lévy intensities.

3 Transport Distances

In this section we define a class of transport distances on CRVs. These are built on the Wasserstein distance [25], whose geometric definition makes it an ideal choice for measuring the similarity between distributions.

Let $\|\cdot\|_m$ denote the Euclidean distance on $\mathbb{R}^m$ and let $\mathbb{N}^+ = \mathbb{N} \setminus \{0\}$. For any pair π_1, π_2 of probability measures on $(\mathbb{R}^m, \|\cdot\|_m)$, we indicate by $C(\pi_1, \pi_2)$ the Fréchet class of π_1 and π_2, i.e. the set of distributions (*couplings*) on the product space $\mathbb{R}^{2m}$ whose marginal distributions coincide with π_1 and π_2 respectively.

Definition 3.1 The Wasserstein distance of order $p \in \mathbb{N}^+$ on $(\mathbb{R}^m, \|\cdot\|_m)$ is defined as

$$\mathcal{W}_p(\pi_1, \pi_2) = \inf_{(Z_1, Z_2) \in C(\pi_1, \pi_2)} \{\mathbb{E}(\|Z_1 - Z_2\|_m^p)\}^{\frac{1}{p}}.$$

By extension, we refer to the Wasserstein distance between two random vectors X_1, X_2 on $\mathbb{R}^m$ as the Wasserstein distance between their laws, i.e. $\mathcal{W}_p(X_1, X_2) = \mathcal{W}_p(\mathcal{L}(X_1), \mathcal{L}(X_2))$.

In the next proposition we show how the Wasserstein distance may be used to define a distance between CRVs in a natural way. The proof is a straightforward generalization of results in [3]. Before providing the main statement, we underline that a CRV has finite moments up to order $p \in \mathbb{N}^+$ if for every $\ell \in \{1, \dots, p\}$,

$$\int_{\mathbb{R}_+^m \times \mathbb{X}} s^\ell\, \nu(ds_1, \dots, ds_m, dx) < +\infty, \tag{3.1}$$

where $s^\ell = (s_1^\ell, \ldots, s_m^\ell)$ and $+\infty = (+\infty, \cdots, +\infty)$. Denote by $\mathbb{P}_p(M_{\mathbb{X}}^m) = \{\mathcal{L}(\tilde{\mu})$ s.t. $\tilde{\mu}$ is a CRV that satisfies (3.1)$\}$.

Proposition 3.2 *For every $p \in \mathbb{N}^+$, the following function $d_{\mathcal{W},p} : \mathbb{P}_p(M_{\mathbb{X}}^m) \times \mathbb{P}_p(M_{\mathbb{X}}^m) \to [0, +\infty)$ defines a distance:*

$$d_{\mathcal{W},p}(\mathcal{L}(\tilde{\mu}_1), \mathcal{L}(\tilde{\mu}_2)) = \sup_{A \in \mathcal{X}} W_p(\tilde{\mu}_1(A), \tilde{\mu}_2(A)). \tag{3.2}$$

By extension, we refer to the distance $d_{\mathcal{W},p}$ between CRVs as the distance between their laws. The natural definition of $d_{\mathcal{W},p}$ is often coupled with analytical tractability, as shown in [2] and [3], which makes it particularly attractive in a number of statistical applications. In particular, in [2], $d_{\mathcal{W},1}$ was used in the one-dimensional scenario, i.e. between the laws of completely random measures, to measure the discrepancy between Bayesian nonparametric models for exchangeable time-to-event data. On the other hand, in [3], $d_{\mathcal{W},2}$ was used to measure the dependence structure of a CRV.

4 Measuring Dependence in Bayesian Nonparametrics

In the last 20 years Bayesian nonparametric models have gone beyond the exchangeability assumption through the introduction of dependent random measures, which provide a flexible framework for modeling the heterogeneity across multiple populations. The prior dependence between random measures regulates the borrowing of strength across different populations and thus needs a careful elicitation. The current state-of-the-art is to provide the analytical expression for the linear correlation $Corr(\tilde{\mu}_1(A), \tilde{\mu}_2(A))$, which only captures partial information about the dependence structure. In [3] the authors propose to use the distance defined in Proposition 3.2 to compare different dependence structures between CRVs with equal marginal distributions, i.e. in the same Fréchet class. In particular, one may define an overall measure of dependence of $\tilde{\mu}$ by considering its distance from the maximally dependent CRV in the same Fréchet class, usually referred to as the comonotonic vector $\tilde{\mu}^{\mathrm{co}}$:

$$\mathrm{Dep}(\tilde{\mu}) = d_{\mathcal{W},2}(\tilde{\mu}, \tilde{\mu}^{\mathrm{co}}) \tag{4.1}$$

The goal is to find tight bounds for $\mathrm{Dep}(\tilde{\mu})$ in terms of the hyperparameters of the model, in order to quantify their impact on the dependence structure for a principled prior elicitation. This is achieved in [3] by (1) using compound Poisson approximations; (2) finding a new upper bound on the Wasserstein distance between bivariate compound Poisson distributions; (3) finding the expression for the optimal coupling between a distribution on $\mathbb{R}^2$ and the comonotonic one in the same Fréchet class. In particular, one finds tight upper bounds for $\mathrm{Dep}(\tilde{\mu})$ in terms of the underlying bivariate Lévy measures. This allows to treat many noteworthy dependence structures, such as GM-dependence [11, 16, 17], compound random measures [10, 23] and Clayton Lévy copulae [4, 8, 15, 24].

5 Survival Analysis in Bayesian Nonparametrics

Completely random measures play a particularly important role in Bayesian nonparametric models for time-to-event data. Here the main quantities of interest are typically the survival function, cumulative hazards and the hazard function. Consequently, the most notable Bayesian nonparametric models in survival analysis and reliability theory provide closed form estimates of these functions, in terms of underlying CRMs. Among the models for the hazard function, the one proposed by Dykstra and Laud [7] stands out for combining both flexibility and tractability. The random hazard function $\tilde{h}$ is modeled as a kernel mixture over a CRM:

$$\tilde{h}(t) = \int_{\mathbb{X}} k(t|x)\, \tilde{\mu}(dx), \tag{5.1}$$

where $t \in \mathbb{R}^+$, $k : \mathbb{X} \times \mathbb{R}^+ \to \mathbb{R}^+$ is a measurable kernel and $\tilde{\mu}$ is a CRM on $\mathbb{X}$ with Lévy intensity ν. The original model in [7] was defined for $k(t|x) = \beta(x)\mathbb{1}_{(0,t]}(x)$, where $\beta : \mathbb{R}^+ \to \mathbb{R}^+$ is a measurable function, and $\tilde{\mu}$ a gamma CRM, i.e. such that the Lévy intensity satisfies

$$\nu(ds, dx) = \frac{e^{-bs}}{s}\mathbb{1}_{(0,+\infty)}(s)\, ds\, \alpha(dx), \tag{5.2}$$

for $b > 0$ and $\alpha \in M_{\mathbb{X}}$. We write $\tilde{\mu} \sim \mathrm{Ga}(b, \alpha)$. The extension to a general kernel was proposed in [19]. Later work by James [12] also extended this model to a general CRM.

The hazard model (5.1) is very flexible and incorporates a wide variety of distributional assumptions. On the other hand, it is not easy to understand how the parameters of the CRM and the kernel function impact the distribution of the random hazard. For a careful prior elicitation and sensitivity analysis, a principled quantification of the discrepancy at the level of the hazards is of fundamental need. We choose the Wasserstein distance of order 1 as a measure of discrepancy and we seek for an analytical expression of

$$\sup_{t \in [0,T]} \mathcal{W}_1(\tilde{h}_1(t), \tilde{h}_2(t)) \tag{5.3}$$

where $T > 0$ and $\tilde{h}_1, \tilde{h}_2$ are two different specifications for (5.1). We point out that $[0, T]$ may be interpreted as a time interval of interest, typically coinciding with the start and the end of the study. The analytical evaluation of a distance is a difficult task in general, even more so since the law of the random variable in (5.1) is defined in an indirect way through the Lévy measure of the mixing CRM. Nonetheless, in [2] the authors were able to find informative bounds on the distance in (3.2) in terms of the corresponding Lévy measures by (1) leveraging compound Poisson approximations of the completely random measures; (2) bounding the Wasserstein distance between compound Poisson distributions, as first suggested by Mariucci and Reiß [20] in the context of Lévy processes. Then, the deep connections between (5.3) and (3.2) lead to the following theorem, whose proof may be found in [2]. Before providing the main statement, we stress two conditions on the kernels.

$$\lim_{t\to\infty} \int_0^t \int_{\mathbb{R}^+\times\mathbb{X}} k(u\,|\,x)\, s\, du\, \mathcal{N}(ds, dx) = +\infty, \tag{5.4}$$

$$\int_{\mathbb{R}^+\times\mathbb{X}} k(t|x)\, s\, \nu(ds, dx) < +\infty. \tag{5.5}$$

Theorem 5.1 *Let $\tilde{h}_1 = \{\tilde{h}_1(t)\,|\,t \geq 0\}$ and $\tilde{h}_2 = \{\tilde{h}_2(t)\,|\,t \geq 0\}$ be random hazard rates as in (5.1) with associated infinitely active CRMs $\tilde{\mu}_i$, Lévy intensity ν_i, and kernel k_i that satisfy (5.4) and (5.5), for $i = 1, 2$. Then the Wasserstein distance between the marginal hazard rates is finite and for every $t \geq 0$,*

$$g_{low}(t) \leq \mathcal{W}_1(\tilde{h}_1(t), \tilde{h}_2(t)) \leq g_{up}(t),$$

where

$$g_{low}(t) = \left| \int_{\mathbb{R}^+\times\mathbb{X}} k_1(t|x)\, s\, \nu_1(ds, dx) - \int_{\mathbb{R}^+\times\mathbb{X}} k_2(t|x)\, s\, \nu_2(ds, dx) \right|,$$

$$g_{up}(t) = \int_0^{+\infty} \left| \int_{(u,+\infty)\times\mathbb{X}} \frac{1}{k_1(t\,|\,x)} \nu_1\Big(d\frac{s}{k_1(t\,|\,x)}, dx\Big) - \frac{1}{k_2(t\,|\,x)} \nu_2\Big(d\frac{s}{k_2(t\,|\,x)}, dx\Big) \right| du.$$

In particular if there exists a dominating measure η such that the Radon–Nikodym derivatives $\nu_i(s, x)$ satisfy, for $i \neq j$ in $\{1, 2\}$,

$$\frac{1}{k_i(t|x)} \nu_i\Big(\frac{s}{k_i(t|x)}, x\Big) \leq \frac{1}{k_j(t|x)} \nu_j\Big(\frac{s}{k_j(t|x)}, x\Big) \tag{5.6}$$

for all $(s, x) \in \mathbb{R}^+ \times \mathbb{X}$, then

$$\mathcal{W}_1(\tilde{h}_1(t), \tilde{h}_2(t)) = \left| \int_{\mathbb{R}^+\times\mathbb{X}} k_1(t|x)\, s\, \nu_1(ds, dx) - \int_{\mathbb{R}^+\times\mathbb{X}} k_2(t|x)\, s\, \nu_2(ds, dx) \right|.$$

Theorem 5.1 was used in [2] to measure the discrepancy between hazards with kernels of the type of [7], i.e. $k(t|x) = \beta(x)\mathbb{1}_{[0,t]}(x)$, which is a popular choice when modeling increasing hazards, in the particular case where $\beta(x) = \beta$ is a constant function. This specification is very common in applications and brings to the following measurement of discrepancy, whose proof may be found in [2]. We denote by Leb^+ the Lebesgue measure on $(0, +\infty)$.

Theorem 5.2 *Let $\tilde{\mu}_i \sim \mathrm{Ga}(b_i, \mathrm{Leb}^+)$ as defined in (5.2) and let $k_i(t|x) = \beta_i \mathbb{1}_{[0,t]}(x)$, with $b_i, \beta_i > 0$, for $i = 1, 2$. If $\tilde{h}_1$ and $\tilde{h}_2$ are the corresponding hazard rate mixtures, then*

$$\mathcal{W}_1(\tilde{h}_1(t), \tilde{h}_2(t)) = t\left|\frac{\beta_1}{b_1} - \frac{\beta_2}{b_2}\right|.$$

6 Time-Dependent Kernels

In this section we make some progress in the understanding of the distributional implications of the hazard rate model in (5.1) when $\tilde{\mu} \sim \mathrm{Ga}(b, \mathrm{Leb}^+)$ as defined in (5.2). In particular, the goal is to understand the impact of using a kernel of the type of [7] when the time influences also the functional form of the kernel and not only the support, i.e. $\beta(\cdot)$ is not constant in $(0, t]$. This scenario is of particular importance when we judge that the index of dispersion varies in time, since when $\beta(x) = \beta > 0$,

$$\frac{\mathrm{Var}(\tilde{h}(t))}{\mathbb{E}(\tilde{h}(t))} = \frac{\beta}{b}.$$

We thus consider the scenario where $\beta(x) = \beta + \gamma x$, with $\beta, \gamma > 0$.

Theorem 6.1 *Let $\tilde{\mu}_i \sim \mathrm{Ga}(b_i, \mathrm{Leb}^+)$ as defined in (5.2) and let $k_1(t|x) = \beta\mathbb{1}_{[0,t]}(x)$ and $k_2(t|x) = (\beta + \gamma x)\mathbb{1}_{[0,t]}(x)$, with $b_1, b_2, \beta, \gamma > 0$. If $\tilde{h}_1$ and $\tilde{h}_2$ are the corresponding hazard rate mixtures, then*

1. *If $b_1 \geq b_2$,*

$$\mathcal{W}_1(\tilde{h}_1(t), \tilde{h}_2(t)) = \left(\frac{1}{b_2} - \frac{1}{b_1}\right)\beta t + \frac{\gamma}{2b_2}t^2.$$

2. *If $b_1 \leq b_2$ and $t \leq \beta(b_2 - b_1)/(b_1\gamma)$*

$$\mathcal{W}_1(\tilde{h}_1(t), \tilde{h}_2(t)) = \left(\frac{1}{b_1} - \frac{1}{b_2}\right)\beta t - \frac{\gamma}{2b_2}t^2.$$

3. *Otherwise,*

$$g_{low}(t) \leq \mathcal{W}_1(\tilde{h}_1(t), \tilde{h}_2(t)) \leq g_{up}(t),$$

where

$$g_{low}(t) = \left(\frac{1}{b_2} - \frac{1}{b_1}\right)\beta t + \frac{\gamma}{2b_2}t^2$$

$$g_{up}(t) = \left(\frac{1}{b_2} - \frac{1}{b_1}\right)^2 \frac{\beta^2 b_2}{\gamma} + \left(\frac{1}{b_2} - \frac{1}{b_1}\right)\beta t + \frac{\gamma}{2b_2}t^2$$

Proof First of all we observe that

$$\nu_{k,1}(ds, dx) = \frac{1}{k_1(t|x)}\nu_1\left(d\frac{s}{k_1(t|x)}\,dx\right) = \frac{e^{-\frac{s\,b_1}{\beta}}}{s}\,\mathbb{1}_{(0,+\infty)}(s)\,\mathbb{1}_{[0,t]}(x)\,ds\,dx,$$

$$\nu_{k,2}(ds, dx) = \frac{1}{k_2(t\,|\,x)}\nu_2\left(d\frac{s}{k_2(t|x)}\,dx\right) = \frac{e^{-\frac{s\,b_2}{\beta+\gamma x}}}{s}\,\mathbb{1}_{(0,+\infty)}(s)\,\mathbb{1}_{[0,t]}(x)\,ds\,dx.$$

Since $\gamma > 0$, (5.6) holds whenever $b_1 \geq b_2$. Part 1 of the statement thus holds by Theorem 5.1, by observing that

$$\int_{\mathbb{R}^+\times\mathbb{R}} k_2(t\,|\,x)\,s\,\nu_2(ds, dx) = \frac{\beta}{b_2}t + \frac{\gamma}{2b_2}t^2.$$

As for part 2 of the statement, it suffices to prove the expression for the upper bound. With a slight abuse of notation, indicate by $\nu_{k,i}(s, x)$ the Radon–Nikodym derivative of $\nu_{k,i}(ds, dx)$, for $i = 1, 2$. We observe that $\nu_{k,1}(s, x) \leq \nu_{k,2}(s, x)$ for every $s > 0$ and every $t \geq y \geq \beta(b_2 - b_1)/(b_1\gamma)$, so that the Radon–Nikodym derivatives are not globally ordered. We denote by $\delta = \min(\beta(b_2 - b_1)/(b_1\gamma), t)$. We then have

$$\begin{aligned}
&\int_0^{+\infty}\left|\int_{(u,+\infty)\times\mathbb{R}} (\nu_{k,1}(s, x) - \nu_{k,2}(s, x))\,ds\,dx\right| du \\
&\leq \int_0^{+\infty}\int_{(u,+\infty)}\int_{\mathbb{R}} |\nu_{k,1}(s, x) - \nu_{k,2}(s, x)|\,dx\,ds\,du.
\end{aligned}$$

By interchanging the integrals thanks to Fubini's Theorem, this is equal to

$$\begin{aligned}
&\int_0^{+\infty}\int_{\mathbb{R}}\left(\int_0^s du\right)|\nu_{k,1}(s, x) - \nu_{k,2}(s, x)|\,dx\,ds \\
&= \int_0^{+\infty}\int_{\mathbb{R}} |s\nu_{k,1}(s, x) - s\nu_{k,2}(s, x)|\,dx\,ds \\
&= \int_0^{+\infty}\left(\int_0^{\delta}(e^{-\frac{s\,b_1}{\beta}} - e^{-\frac{s\,b_2}{\beta+\gamma x}})\,dx + \int_{\delta}^{t}(e^{-\frac{s\,b_2}{\beta+\gamma x}} - e^{-\frac{s\,b_1}{\beta}})\,dx\right)ds
\end{aligned}$$

$$= \int_0^\delta \left(\frac{\beta}{b_1} - \frac{\beta + \gamma x}{b_2} \right) dx + \int_\delta^t \left(\frac{\beta + \gamma x}{b_2} - \frac{\beta}{b_1} \right) dx.$$

The conclusion follows by simple calculations. □

Theorem 6.1 allows one to measure the impact of introducing a time dependent function in the kernel of [7]. In particular, we underline how the discrepancy grows quadratically in time, thus greatly influencing our prior opinion on the process for large t. Moreover, as $t \to +\infty$ we observe that the upper and lower bounds are asymptotically equivalent, providing the exact leading term for the Wasserstein distance.

7 Discussion and Further Work

In this paper we have discussed two different frameworks where transport distances between random vectors of measures provide deeper insights on notable Bayesian nonparametric models, favoring the elicitation of the prior. The amount of dependence in partially exchangeable models regulates the borrowing of information across groups, with a large impact on the inference. It is thus of fundamental importance to translate our prior beliefs on the dependence structure into the specification of the prior. Since exchangeability corresponds to a situation of maximal dependence, it seems natural to encode the prior beliefs on the dependence structure in terms of distance from the exchangeable scenario, as in (4.1). By choosing a subjective threshold τ for such distance, the tight upper bounds found in [3] may be set to be equal to τ by choosing appropriate values of the hyperparameters of the model. Moreover, Theorem 6.1 may be used for the prior elicitation of hazard rate models as in (5.1). The kernel $k(t|x) = \beta \mathbb{1}_{[0,t]}(x)$ with $\beta > 0$ is the most common specification in applications involving increasing hazard rates and may be treated as a reference kernel. However, if one believes that the index of dispersion varies over the time interval of interest $(0, t^*]$, it is natural to use a time varying specification as $k(t|x) = (\beta + \gamma x)\mathbb{1}_{[0,t]}(x)$, though securing a certain degree of similarity with respect to the reference kernel for every $t \in (0, t^*]$. By measuring the similarity in terms of Wasserstein distance and fixing a subjective threshold τ, the exact value of the distance or its upper bound in Theorem 6.1 is maximized in t^*. One can then set the upper bound at time t^* equal to τ, so that the hyperparameter γ may be chosen and elicited accordingly.

Completely random measures are widely used because they combine modeling flexibility with analytical tractability. In particular, there are many closed form results for the posterior distribution of the random measures given exchangeable or partially exchangeable observations. These have been used for example in [2] to evaluate approximation errors of a posterior sampling scheme in terms of the Wasserstein distance. Future research will concern the analysis of the dependence structure of the posterior distribution $\tilde{\mu}^*$ through $\mathrm{Dep}(\tilde{\mu}^*)$ in (4.1). The plan would then be to use this to test whether the data supports the heterogeneity assumption across groups, along the lines of the parametric tests developed in [1].

Acknowledgement Antonio Lijoi and Igor Prünster are partially supported by MIUR, PRIN Project 2015SNS29B.

References

1. Bacallado, S., Diaconis, P., Holmes, S.: de Finetti priors using Markov chain Monte Carlo computations. J. Stat. Comput. **25**, 797–808 (2015)
2. Catalano, M., Lijoi, A., Prünster, I.: Approximation of Bayesian models for time-to-event data. Electron. J. Stat. **14**, 3366–3395 (2020)
3. Catalano, M., Lijoi, A., Prünster, I.: Measuring the dependence in the Wasserstein distance for Bayesian nonparametric models. Annals of Statistics, forthcoming. DOI: https://doi.org/10.1214/21-AOS2O6S (2021)
4. Cont, R., Tankov, P.: Financial Modeling with Jump Processes. Chapman and Hall/CRC, Boca Raton (2004)
5. Daley, D.J., Vere-Jones, D.: An Introduction to the Theory of Point Processes: Volume I: Elementary Theory and Methods. Probability and Its Applications. Springer, Berlin (2002)
6. Doksum, K.: Tailfree and neutral random probabilities and their posterior distributions. Ann. Probab. **2**, 183–201 (1974)
7. Dykstra, R.L., Laud, P.: A Bayesian nonparametric approach to reliability. Ann. Stat. **9**, 356–367 (1981)
8. Epifani, I., Lijoi, A.: Nonparametric priors for vectors of survival functions. Stat. Sin. **20**, 1455–1484 (2010)
9. Ferguson, T.S.: Bayesian density estimation by mixtures of normal distributions. In: Recent Advances in Statistics, pp. 287–302. Academic Press, New York (1983)
10. Griffin, J.E., Leisen, F.: Compound random measures and their use in Bayesian nonparametrics. JRSS B **79**, 525–545 (2017)
11. Griffiths, R., Milne, K.R.: A class of bivariate Poisson processes. J. Multivar. Anal. **8**, 380–395 (1978)
12. James, L.F.: Bayesian Poisson process partition calculus with an application to Bayesian Lévy moving averages. Ann. Stat. **33**, 1771–1799 (2005)
13. Kallenberg, O.: Random Measures, Theory and Applications. Springer International Publishing, Cham (2017)
14. Kingman, J.F.C.: Completely random measures. Pacific J. Math. **21**, 59–78 (1967)
15. Leisen, F., Lijoi, A.: Vectors of two-parameter Poisson-Dirichlet processes. J. Multivar. Anal. **102**, 482–495 (2011)
16. Lijoi, A., Nipoti, B.I.: A class of hazard rate mixtures for combining survival data from different experiments. J. Am. Stat. Assoc. **20**. 802–814 (2014)
17. Lijoi, A., Nipoti, B., Prünster, I.: Bayesian inference with dependent normalized completely random measures. Bernoulli **20**, 1260–1291 (2014)
18. Lo, A.: On a class of Bayesian nonparametric estimates: I. Density estimates. Ann. Stat. **12**, 351–357 (1984)
19. Lo, A., Weng, C.: On a class of Bayesian nonparametric estimates: II. Hazard rate estimates. Ann. Inst. Stat. Math. **41**, 227–245 (1989)
20. Mariucci, E., Reiß, M.: Wasserstein and total variation distance between marginals of Lévy processes. Electron. J. Stat. **12**, 2482–2514 (2018)
21. Quintana, F.A., Müller, P., Jara, A., MacEachern, S.N.: The dependent Dirichlet process and related models. arXiv 2007.06129 (2020)

22. Regazzini, E., Lijoi, A., Prünster, I.: Distributional results for means of normalized random measures with independent increments. Ann. Stat. **31**, 560–585 (2003)
23. Riva–Palacio, A., Leisen, F.: Compound vectors of subordinators and their associated positive Lévy copulas. arXiv 1909.12112 (2019)
24. Tankov, P.: Dependence structure of spectrally positive multidimensional Lévy processes. Unpublished manuscript (2003)
25. Villani, C.: Optimal Transport: Old and New. Springer, Berlin Heidelberg (2008)

Bayesian Non-parametric Priors Based on Random Sets

María F. Gil-Leyva

Abstract We study the construction of random discrete distributions, taking values in the infinite dimensional simplex, by means of a latent random subset of the natural numbers. The derived sequences of random weights are then used to establish a Bayesian non-parametric prior. A sufficient condition on the distribution of the random set is given, that assures the corresponding prior has full support, and taking advantage of the construction, we propose a general MCMC algorithm for density estimation purposes. This method is illustrated by building a new distribution over the space of all finite and non-empty subsets of $\mathbb{N}$, that subsequently leads to a general class of random probability measures termed Geometric product stick-breaking process. It is shown that Geometric product stick-breaking process approximate, in distribution, Dirichlet and Geometric processes, and that the respective weights sequences have heavy tails, thus leading to very flexible mixture models.

Keywords Bayesian non-parametric prior · Density estimation · Dirichlet process · Geometric process · Random sets

1 Introduction

The study of distributions over the infinite dimensional simplex, Δ_∞, arises naturally in the theory of Bayesian non-parametric statistics, as they allow the construction of mathematically tractable random probability measures, that are latter assumed to drive some exchangeable sequence. Namely, given a sequence of weights, $\mathbf{W} = (\mathbf{w}_j)_{j\geq 1}$, taking values in $\Delta_\infty = \{(w_j)_{j\geq 1} \in [0,1]^\infty : w_j \geq 0, \sum_{j\geq 1} w_j = 1\}$, an independent sequence $\boldsymbol{\Xi} = (\boldsymbol{\xi}_j)_{j\geq 1} \overset{\text{iid}}{\sim} P_0$, for some diffuse

M. F. Gil-Leyva (✉)
IIMAS, Universidad Nacional Autónoma de México, CDMX, Mexico
e-mail: marifer@sigma.iimas.unam.mx

D. Hernández-Hernández et al. (eds.), *Advances in Probability and Mathematical Statistics*, Progress in Probability 79, https://doi.org/10.1007/978-3-030-85325-9_5

distribution, P_0 over a Polish space $(\mathcal{S}, \mathcal{B}(\mathcal{S}))$, and probability kernel,[1] ψ, from $\mathcal{S}$ into the Polish space $(\mathcal{R}, \mathcal{B}(\mathcal{R}))$, one can construct the random mixture,

$$\mathbf{p} = \sum_{j\geq 1} \mathbf{w}_j \psi(\cdot|\boldsymbol{\xi}_j), \tag{1.1}$$

and given $\mathbf{p}$, model data, $\{\mathbf{y}_1, \mathbf{y}_2, \ldots\}$, as i.i.d. sampled from $\mathbf{p}$ ($\{\mathbf{y}_1, \mathbf{y}_2, .. \mid \mathbf{p}\} \overset{\text{iid}}{\sim} \mathbf{p}$). This is generally interpreted as if $\mathbf{y}_k$ is sampled from the distribution $\psi(\cdot|\boldsymbol{\xi}_j)$, with probability $\mathbf{w}_j$, for every $j \geq 1$. In this context two widely studied and useful subclasses of random mixtures are the ones that allow a density and the class of species sampling processes. The first class arises when for each $s \in \mathcal{S}$, $\psi(\cdot|s)$ has a density, $f(\cdot|s)$ with respect to a suitable diffuse measure. This way one can consider the random mixture of densities $\sum_{j\geq 1} \mathbf{w}_j f(\cdot|\boldsymbol{\xi}_j)$, and work with this one instead of $\mathbf{p}$. The second class emerges if we fix $\mathcal{S} = \mathcal{R}$ and $\psi(\cdot|\boldsymbol{\xi}_j) = \delta_{\boldsymbol{\xi}_j}$, then $\mathbf{p}$ becomes a (proper) species sampling process [17]. While mixtures of random densities describe random probability measures that are diffuse almost surely, proper species sampling processes are purely atomic almost surely. Arbitrary random mixtures, can be constructed by means of measurably transforming a proper sampling process $\sum_{j\geq 1} \mathbf{w}_j \delta_{\boldsymbol{\xi}_j}$, through

$$\sum_{j\geq 1} \mathbf{w}_j \delta_{\boldsymbol{\xi}_j} \mapsto \int_{\mathcal{S}} \psi(\cdot|s)\mu(ds) = \sum_{j\geq 1} \mathbf{w}_j\, \psi(\cdot|\boldsymbol{\xi}_j).$$

For this reason, the distribution of the (latent) species sampling processes, in a Bayesian context, is referred to as the prior distribution of the model. Evidently, the prior distribution is completely characterized by choosing the diffuse probability measure, P_0, and defining the distribution of the weights, $\mathbf{W}$, which is the most challenging part. The canonical example of Bayesian non-parametric priors is the Dirichlet process [3, 5, 19], searching for generalizations and competitive alternatives to the canonical model, various techniques to place distributions on Δ_∞ have been developed. Some of the most notable are through the normalization of the jumps of a subordinator (e.g. [9, 11, 13, 18]), through a prediction rule (e.g. [3, 16, 17]), and by means of the stick-breaking construction of a weights sequence (e.g. [10, 14, 19]). The latter consists in decomposing the weight as

$$\mathbf{w}_1 = \mathbf{v}_1, \quad \mathbf{w}_j = \mathbf{v}_j \prod_{i=1}^{j-1}(1 - \mathbf{v}_i), \quad j \geq 2, \tag{1.2}$$

[1] Recall that a probability kernel ψ from $\mathcal{S}$ into $(\mathcal{R}, \mathcal{B}(\mathcal{R}))$ is a function, $\psi : \mathcal{B}(\mathcal{R}) \times \mathcal{S} \to [0, 1]$, such that for every $s \in \mathcal{S}$ fixed $\psi(\cdot|s)$ is a probability measure, and for every $B \in \mathcal{B}(\mathcal{R})$ fixed, $\psi(B|\cdot)$ is a measurable with respect to $\mathcal{B}(\mathcal{S})$ and $\mathcal{B}([0, 1])$.

for some sequence taking values in $[0, 1]$, $\mathbf{V} = (\mathbf{v}_i)_{i\geq 1}$, hereinafter referred to as length variables. Particularly, if $(\mathbf{v}_i)_{i\geq 1} \overset{\text{iid}}{\sim} \mathsf{Be}(1, \theta)$, for some $\theta > 0$, the Dirichlet processes [3, 5, 19] arises. Another well-known stick-breaking process is the Geometric process [7], which has length variables $\mathbf{v}_i = \boldsymbol{\lambda} \sim \mathsf{Be}(\alpha, \beta)$ for every $i \geq 1$, so that (1.2) reduces to $\mathbf{w}_j = \boldsymbol{\lambda}(1 - \boldsymbol{\lambda})^{j-1}$, $j \geq 1$. Further examples of Bayesian non parametric priors, random mixtures, related models and their applications can be found in [6, 8, 9].

In spite of the construction method of a random mixture, as explained in [2], an essential requirement to the prior distribution is for it to have full support. This property is determined by the weights, and assures that if the support of P_0 is $\mathcal{S}$, then the support of the prior (with respect to the topology of weak convergence) is the set of all probability measures over $(\mathcal{S}, \mathcal{B}(\mathcal{S}))$. As also shown in [2], the prior has full support if and only if for every $\varepsilon > 0$, $\mathbb{P}[\max_{j\geq 1} \mathbf{w}_j < \varepsilon] > 0$. This immediately discards sequences of weights for which there exists $m \in \mathbb{N}$, such that $\sum_{j=1}^{m} \mathbf{w}_j = 1$ almost surely, as this implies $\mathbf{w}_j > 1/m$, for all $j \geq 1$. In other words, random mixtures depending on arbitrarily many random variables are the most flexible models, however, the same characteristic also makes them challenging to implement. In order to overcome this, Walker in [20] suggested to construct a latent random set, say $\boldsymbol{\Phi}$, that takes values in $\mathcal{F}_{\mathbb{N}} = \{A \subseteq \mathbb{N} : 0 < |A| < \infty\}$, and that makes sampling from $\mathbf{p}$ equivalent to sampling from the uniform finite mixture $|\boldsymbol{\Phi}|^{-1} \sum_{j\in\boldsymbol{\Phi}} \psi(\cdot|\boldsymbol{\xi}_j)$. This approach not only solves the practical problem of dealing with infinitely many random variables, but also suggests a new form to construct distributions over Δ_∞.

The rest of the paper is organized as follows. In Sect. 2 we formalize the procedure of how to construct a distribution over Δ_∞ given a parametric distribution, π_{Φ}, over $\mathcal{F}_{\mathbb{N}}$. Inhere we also give sufficient conditions on π_{Φ} to assure the corresponding prior has full support, and propose a general MCMC algorithm for density estimation purposes. In Sect. 3.1 we illustrate the notions analysed in Sect. 2, by reviewing some examples that already appear in the literature, and in Sect. 3.2 we do so by means of a new class of distributions over $\mathcal{F}_{\mathbb{N}}$, characterized by the fact that the elements of corresponding random sets, are chosen by means of independent Bernoulli trials. The fourth and final section, is dedicated to study further properties of the model that arises from the construction in Sect. 3.2, such as the fact that Geometric processes and Dirichlet processes can be recovered for extreme values of the hyper-parameters and that the corresponding weights sequence is heavy-tailed.

2 Distributions on the Simplex Based on Distributions of Random Sets

Consider a random element, $\boldsymbol{\tau}$, taking values in some Polish space $(\mathcal{T}, \mathcal{B}(\mathcal{T}))$ and a mass probability kernel π_{Φ} from $\mathcal{T}$ into $\mathcal{F}_{\mathbb{N}}$, so that for each $t \in \mathcal{T}$, $\pi_{\Phi}(\cdot \mid t)$ is a mass probability function over $\mathcal{F}_{\mathbb{N}}$, and for each $A \in \mathcal{F}_{\mathbb{N}}$, $\pi_{\Phi}(A \mid \cdot)$ is a measurable function from $(\mathcal{T}, \mathcal{B}(\mathcal{T}))$ into $([0,1], \mathcal{B}([0,1]))$. Let $\mathbf{\Phi}$ be some random set satisfying $\{\mathbf{\Phi}|\boldsymbol{\tau}\} \sim \pi_\Phi(\cdot|\boldsymbol{\tau})$, and say that given $\mathbf{\Phi}$, we uniformly pick one of its elements, $\mathbf{d}$. Define $\mathbf{w}_j$ as the conditional probability that $\mathbf{d} = j$ given $\boldsymbol{\tau}$, that is $\mathbf{w}_j = \mathbb{P}[\mathbf{d} = j \mid \boldsymbol{\tau}]$. Under the assumption that $\mathbf{d}$ is conditionally independent of $\boldsymbol{\tau}$ given $\mathbf{\Phi}$, by the tower property of conditional expectation we obtain,

$$\mathbf{w}_j = \mathbb{E}\left[\frac{1}{|\mathbf{\Phi}|}\mathbf{1}_{\{j\in\mathbf{\Phi}\}}\,\middle|\,\boldsymbol{\tau}\right] = \sum_{A\in\mathcal{F}_{\mathbb{N}}} \frac{1}{|A|}\mathbf{1}_{\{j\in A\}}\pi_\Phi(A \mid \boldsymbol{\tau}), \quad j \geq 1. \tag{2.1}$$

Insomuch as the events $(\mathbf{d} = j)_{j\geq 1}$ are mutually disjoint and its union $(\mathbf{d} \in \mathbb{N})$ occurs almost surely we must have $\sum_{j\geq 1}\mathbf{w}_j = 1$. In other words, by conditional monotone convergence theorem, and since $\sum_{j\geq 1}\mathbf{1}_{\{j\in\mathbf{\Phi}\}} = |\mathbf{\Phi}|$, we get

$$\sum_{j\geq 1}\mathbf{w}_j = \mathbb{E}\left[\frac{1}{|\mathbf{\Phi}|}\sum_{j\geq 1}\mathbf{1}_{\{j\in\mathbf{\Phi}\}}\,\middle|\,\boldsymbol{\tau}\right] = 1, \tag{2.2}$$

almost surely. Thus, each parametric distribution, π_Φ, over $\mathcal{F}_{\mathbb{N}}$, together with a randomization of its parameters, $\boldsymbol{\tau}$, and through (2.1), characterizes completely the law of $\mathbf{W} = (\mathbf{w}_j)_{j\geq 1}$, which takes values in the infinite dimensional simplex. In other words, if we denote by π_τ to the distribution of $\boldsymbol{\tau}$, the pair (π_Φ, π_τ), defines a distribution, say $\pi_{\mathbf{W}}$, over Δ_∞.

A major advantage of constructing distributions over Δ_∞ through distributions over $\mathcal{F}_{\mathbb{N}}$, becomes evident when one is interested in sampling from (or infer about) a random mixture, $\mathbf{p} = \sum_{j\geq 1}\mathbf{w}_j\,\psi(\cdot \mid \boldsymbol{\xi}_j)$, where the weights are as in (2.1). For instance, if one is interested in sampling $\mathbf{d}$ from $\mathbb{P}[\mathbf{d} = j \mid \mathbf{W}] = \mathbf{w}_j$, one can alternatively, first sample from the latent random set $\mathbf{\Phi}$, and then uniformly choose an element of $\mathbf{\Phi}$. In terms of the mixture, if one is able to sample to from $\boldsymbol{\xi}_j \sim P_0$, $\boldsymbol{\tau} \sim \pi_\tau$ and $\{\mathbf{\Phi} \mid \boldsymbol{\tau}\} \sim \pi_\Phi(\cdot \mid \boldsymbol{\tau})$ then a sample from $\{\mathbf{y} \mid \mathbf{\Phi}, (\boldsymbol{\xi}_j)_{j\in\mathbf{\Phi}}\} \sim |\mathbf{\Phi}_k|^{-1}\sum_{j\geq 1}\psi(\cdot \mid \boldsymbol{\xi}_j)$ is also a sample from $\{\mathbf{y} \mid \mathbf{W}, \mathbf{\Xi}\} \sim \mathbf{p}$. The following result formalizes this notion in a slightly more general scenario.

Proposition 2.1 *Let $(\mathcal{T}, \mathcal{B}(\mathcal{T}))$ be a Polish space and consider a mass probability kernel π_Φ from $\mathcal{T}$ to $\mathcal{F}_{\mathbb{N}}$. Let $\boldsymbol{\tau}$ be a random element taking values in $\mathcal{T}$, consider $\{\mathbf{\Phi}_1, \ldots, \mathbf{\Phi}_n \mid \boldsymbol{\tau}\} \overset{iid}{\sim} \pi_\Phi(\cdot \mid \boldsymbol{\tau})$ and define $\mathbf{W} = (\mathbf{w}_j)_{j\geq 1}$ as in (2.1). Also consider an independent collection $\mathbf{\Xi} = (\boldsymbol{\xi}_j)_{j\geq 1}$ taking values in the Polish space $(\mathcal{S}, \mathcal{B}(\mathcal{S}))$ and a probability kernel ψ from S into the Polish space $(\mathcal{R}, \mathcal{B}(\mathcal{R}))$. Say that $\{\mathbf{y}_1, \ldots, \mathbf{y}_n\}$ are conditionally independent given $\mathbf{\Xi}, \mathbf{\Phi}_1, \ldots, \mathbf{\Phi}_n$, with $\{\mathbf{y}_k \mid \mathbf{\Phi}_k, \mathbf{\Xi}\} \sim |\mathbf{\Phi}_k|^{-1}\sum_{j\in\mathbf{\Phi}_k}\psi(\cdot \mid \boldsymbol{\xi}_j)$, and also assume that, $\mathbf{y}_k$ is conditionally independent of $\boldsymbol{\tau}$, given $\mathbf{\Phi}_k$, for $1 \leq k \leq n$. Then $\{\mathbf{y}_1, \ldots, \mathbf{y}_n \mid \mathbf{W}, \mathbf{\Xi}\} \overset{iid}{\sim} \sum_{j\geq 1}\mathbf{w}_j\,\psi(\cdot \mid \boldsymbol{\xi}_j)$.*

Proof Fix $B_1, \dots, B_n \in \mathcal{B}(\mathcal{R})$. By the tower property of conditional expectation and conditional monotone convergence theorem we obtain

$$
\begin{aligned}
\mathbb{P}\left[\bigcap_{k=1}^{n}(y_k \in B_k)\,\middle|\,\boldsymbol{\tau}, \boldsymbol{\Xi}\right] &= \mathbb{E}\left[\mathbb{P}\left[\bigcap_{k=1}^{n}(y_k \in B_k)\,\middle|\,\boldsymbol{\tau}, \boldsymbol{\Xi}, \boldsymbol{\Phi}_1, \dots \boldsymbol{\Phi}_n\right]\,\middle|\,\boldsymbol{\tau}, \boldsymbol{\Xi}\right] \\
&= \mathbb{E}\left[\prod_{k=1}^{n}\left(\frac{1}{|\boldsymbol{\Phi}_k|}\sum_{j\in\boldsymbol{\Phi}_k}\psi(B_k \mid \boldsymbol{\xi}_j)\right)\,\middle|\,\boldsymbol{\tau}, \boldsymbol{\Xi}\right] \\
&= \prod_{k=1}^{n}\mathbb{E}\left[\frac{1}{|\boldsymbol{\Phi}_k|}\sum_{j\geq 1}\mathbf{1}_{\{j\in\boldsymbol{\Phi}_k\}}\psi(B_k \mid \boldsymbol{\xi}_j)\,\middle|\,\boldsymbol{\tau}, \boldsymbol{\Xi}\right] \\
&= \prod_{k=1}^{n}\left(\sum_{j\geq 1}\mathbb{E}\left[\frac{1}{|\boldsymbol{\Phi}_k|}\mathbf{1}_{\{j\in\boldsymbol{\Phi}_k\}}\,\middle|\,\boldsymbol{\tau}\right]\psi(B_k \mid \boldsymbol{\xi}_j)\right) \\
&= \prod_{k=1}^{n}\left(\sum_{j\geq 1}\mathbf{w}_j\,\psi(B_k \mid \boldsymbol{\xi}_j)\right)
\end{aligned}
$$

Finally from (2.1) it is evident that $\mathbf{W}$ is $\boldsymbol{\tau}$-measurable, from which we conclude

$$
\mathbb{P}\left[\bigcap_{k=1}^{n}(y_k \in B_k)\,\middle|\,\mathbf{W}, \boldsymbol{\Xi}\right] = \prod_{k=1}^{n}\left(\sum_{j\geq 1}\mathbf{w}_j\,\psi(B_k \mid \boldsymbol{\xi}_j)\right)
$$

□

This said, the next task is to determine under which conditions of π_{Φ} and π_{τ}, does the corresponding prior (the distribution of $\sum_{j\geq 1}\mathbf{w}_j\delta_{\boldsymbol{\xi}_j}$) has full support. To this aim consider $\boldsymbol{\tau}$, $\boldsymbol{\Phi}$ and $\mathbf{W}$ as in (2.1) and define $\mathbf{N} = |\boldsymbol{\Phi}|$. Notice that, for every $\varepsilon > 0$, conditioning on the event $(\mathbf{N} > 1/\varepsilon) = (\mathbf{N}^{-1} < \varepsilon)$, we have that

$$
\mathbf{w}_j = \mathbb{E}\left[\frac{1}{\mathbf{N}}\mathbf{1}_{\{j\in\boldsymbol{\Phi}\}}\,\middle|\,\boldsymbol{\tau}\right] \leq \mathbb{E}\left[\frac{1}{\mathbf{N}}\,\middle|\,\boldsymbol{\tau}\right] < \varepsilon,
$$

for every $j \geq 1$. Hence, $(\mathbf{N} > 1/\varepsilon)$ is contained is the event $(\max_{j\geq 1}\mathbf{w}_j < \varepsilon)$, which means that

$$
\mathbb{P}\left[\max_{j\geq 1}\mathbf{w}_j < \varepsilon\right] \geq \mathbb{P}\left[\mathbf{N} > \frac{1}{\varepsilon}\right]. \tag{2.3}
$$

If we define

$$\pi_{\mathrm{N}}(n \mid t) = \sum_{A \in \mathcal{F}_{\mathrm{N}}} \pi_{\Phi}(A \mid t) \mathbf{1}_{\{|A|=n\}}, \tag{2.4}$$

for every $t \in \mathcal{T}$ and $n \in \mathbb{N}$, so that $\pi_{\mathrm{N}}(\cdot \mid \boldsymbol{\tau})$ is the conditional mass probability function of $\mathbf{N}$ given $\boldsymbol{\tau}$, then we can rewrite (2.3) as

$$\mathbb{P}\left[\max_{j\geq 1} \mathbf{w}_j < \varepsilon\right] \geq \int \sum_{n\geq (1/\varepsilon)} \pi_{\mathrm{N}}(n \mid t)\pi_{\tau}(dt) = \sum_{n\geq (1/\varepsilon)} \int \pi_{\mathrm{N}}(n \mid t)\pi_{\tau}(dt).$$

From these equations, the following results are straight forward.

Proposition 2.2 *Let $(\mathcal{T}, \mathcal{B}(\mathcal{T}))$ be a Polish space and consider a mass probability kernel π_{Φ} from $\mathcal{T}$ to $\mathcal{F}_{\mathbb{N}}$. Let $\boldsymbol{\tau}$ be a random element taking values in $\mathcal{T}$, $\{\boldsymbol{\Phi} \mid \boldsymbol{\tau}\} \sim \pi_{\Phi}(\cdot \mid \boldsymbol{\tau})$, and define $\mathbf{W}$ through (2.1). Let $\mathbf{N} = |\boldsymbol{\Phi}|$ and say that for every $n \in \mathbb{N}$, $\mathbb{P}[\mathbf{N} > n] > 0$. Then for every $\varepsilon > 0$, $\mathbb{P}\left[\max_{j\geq 1} \mathbf{w}_j < \varepsilon\right] > 0$.*

Corollary 2.3 *In the context of Proposition 2.2, define π_{N} through (2.4). If there exist $B \in \mathcal{B}(\mathcal{T})$ such that $\mathbb{P}[\boldsymbol{\tau} \in B] > 0$, and for every $t \in B$ and $m \in \mathbb{N}$, we have that $\sum_{n\geq m} \pi_{\mathrm{N}}(n \mid t) > 0$. Then for every $\varepsilon > 0$, $\mathbb{P}\left[\max_{j\geq 1} \mathbf{w}_j < \varepsilon\right] > 0$.*

Roughly speaking, if $\boldsymbol{\Phi}$ is allowed to contain arbitrarily many random elements, the largest weight can be arbitrarily small. The latter then shows that the corresponding prior has full support and hence is feasible for Bayesian non-parametric inference. In the following we describe a density estimation algorithm for these models via Gibbs sampler implementation.

2.1 Density Estimation Scheme

Say we model elements in $\mathbf{Y} = \{\mathbf{y}_1, \ldots, \mathbf{y}_m\}$ as conditionally i.i.d. sampled from a random mixture $\mathbf{p} = \sum_{j\geq 1} \mathbf{w}_j \psi(\cdot \mid \boldsymbol{\xi})$, where $\mathbf{W}$ satisfies equation (2.1) for some $\boldsymbol{\tau} \sim \pi_{\tau}$ and π_{Φ}. After possibly enlarging the original probability space, we can construct some latent random sets $\{\boldsymbol{\Phi}_1, \ldots \boldsymbol{\Phi}_n \mid \boldsymbol{\tau}\} \stackrel{\text{iid}}{\sim} \pi_{\Phi}(\cdot \mid \boldsymbol{\tau})$ that satisfy the conditions of Proposition 2.1. So that the following are equivalent in terms of the law of $\mathbf{Y}$,

$$\boldsymbol{\Xi} \stackrel{\text{iid}}{\sim} P_0, \quad \mathbf{W} \sim \pi_{\mathrm{W}}, \quad \{\mathbf{y}_1, \mathbf{y}_2, \ldots | \mathbf{W}, \boldsymbol{\Xi}\} \stackrel{\text{iid}}{\sim} \mathbf{p},$$

and

$$\boldsymbol{\Xi} \stackrel{\text{iid}}{\sim} P_0, \quad \boldsymbol{\tau} \sim \pi_{\tau}, \quad \{\boldsymbol{\Phi}_1, \ldots \boldsymbol{\Phi}_n \mid \boldsymbol{\tau}\} \stackrel{\text{iid}}{\sim} \pi_{\Phi}(\cdot \mid \boldsymbol{\tau}), \quad \{\mathbf{y}_k \mid \boldsymbol{\Phi}_k\} \sim \frac{1}{|\boldsymbol{\Phi}_k|} \sum_{j\in \boldsymbol{\Phi}_k} \psi(\cdot \mid \boldsymbol{\xi}_j),$$

for $1 \leq k \leq m$. If we further define the membership random variables, $\mathbf{d}_k = j$ if and only if $\mathbf{y}_k$ is ultimately sampled from $\psi(\cdot \mid \xi_j)$, we get that $\{\mathbf{y}_k \mid \mathbf{\Phi}_k\} \sim |\mathbf{\Phi}_k|^{-1} \sum_{j \in \mathbf{\Phi}_k} \psi(\cdot \mid \xi_j)$ is equivalent to $\{\mathbf{d}_k \mid \mathbf{\Phi}_k\} \sim \mathsf{Unif}(\mathbf{\Phi}_k)$ and $\{\mathbf{y}_k \mid \mathbf{d}_k\} \sim \psi(\cdot \mid \xi_{\mathbf{d}_k})$. Under the assumption that $\psi(\cdot \mid s)$ has a density $f(\cdot \mid s)$ for each s, and taking into account the latent random variables, the likelihood is easily seen to be

$$\pi\left(\{\mathbf{y}_k, \mathbf{d}_k, \mathbf{\Phi}_k\}_{k=1}^m \mid \boldsymbol{\tau}, \Xi\right) = \prod_{k=1}^{n} f(\mathbf{y}_k \mid \xi_{\mathbf{d}_k}) \frac{1}{|\mathbf{\Phi}_k|} \mathbf{1}_{\{\mathbf{d}_k \in \mathbf{\Phi}_k\}} \pi_{\Phi}(\mathbf{\Phi}_k \mid \boldsymbol{\tau}).$$

The full conditional distributions required to update the random elements at each iteration of the Gibbs sampler are given below.

Updating ξ_j for $j \geq 1$

$$\pi\left(\xi_j \mid \ldots\right) \propto p_0(\xi_j) \prod_{k \in D_j} f\left(\mathbf{y}_k | \xi_j\right) \tag{2.5}$$

where $D_j = \{k : \mathbf{d}_k = j\}$, for every $j \geq 1$, and under the assumption that P_0 has a density $\mathbf{p}_0$ with respect to a suitable measure. If p_0 and f form a conjugate pair, the above is easy to sample from.

Updating $\mathbf{d}_k$ for $k \in \{1, \ldots, n\}$

$$\pi\left(\mathbf{d}_k = j \mid \ldots\right) \propto f\left(\mathbf{y}_k \mid \xi_j\right) \mathbf{1}_{\{j \in \mathbf{\Phi}_k\}}, \quad j \geq 1, \tag{2.6}$$

for every $k \in \{1, \ldots, m\}$. Insomuch as $\mathbf{\Phi}_k$ is non-empty and finite a.s., the full conditional of $\mathbf{d}_k$ is a discrete distribution with finite support, hence easy to sample from.

Updating $\mathbf{\Phi}_k$ for $k \in \{1, \ldots, n\}$

$$\pi\left(\mathbf{\Phi}_k = A \mid \ldots\right) \propto \frac{\pi_{\phi}(A \mid \boldsymbol{\tau})}{|A|} \mathbf{1}_{\{\mathbf{d}_k \in A\}}, \quad A \in \mathcal{F}_{\mathbb{N}}, \tag{2.7}$$

for every $k \in \{1, \ldots, m\}$.

Updating $\boldsymbol{\tau}$

$$\pi\left(\boldsymbol{\tau} \mid \ldots\right) \propto \pi_{\tau}(\boldsymbol{\tau}) \prod_{k=1}^{m} \pi_{\phi}(\mathbf{\Phi}_k \mid \boldsymbol{\tau}). \tag{2.8}$$

Evidently the method to sample form Eqs. (2.7) and (2.8) depends on the choice of π_{Φ} and π_{τ}. In the following section we illustrate how to sample from (2.7) and (2.8) for some examples that have already been studied in the literature and a new one. Before moving on we shall mention that when sampling from (2.7) and (2.8) is feasible, given the samples,

$$\left(\left(\mathbf{d}_k^{(i)}\right)_k, \left(\mathbf{\Phi}_k^{(i)}\right)_k, \left(\boldsymbol{\tau}^{(i)}\right)_k, \left(\boldsymbol{\xi}_j^{(i)}\right)_j\right)_{i=1}^{I},$$

obtained after I iterations of the Gibbs sampler, once the burn-in period has elapsed, we estimate the density of the data at y through

$$\mathbb{E}\left[\sum_{j\geq 1} \mathbf{w}_j f(y \mid \boldsymbol{\xi}_j) \,\middle|\, \mathbf{y}_1, \ldots \mathbf{y}_n\right] \approx \frac{1}{I}\sum_{i=1}^{I} \frac{1}{m}\sum_{k=1}^{m} \frac{1}{\left|\mathbf{\Phi}_k^{(i)}\right|} \sum_{j\in\mathbf{\Phi}_k^{(m)}} f\left(y \,\middle|\, \boldsymbol{\xi}_j^{(m)}\right). \tag{2.9}$$

3 Examples

3.1 Random Sets with No Gaps

Consider a random set, $\mathbf{\Phi}$, taking the almost sure form $\mathbf{\Phi} = \{1, \ldots, \mathbf{N}\}$, for some random variable, $\mathbf{N}$, that takes values in $\mathbb{N}$. Evidently $\mathbf{\Phi}$ takes values in $\mathcal{F}_{\mathbb{N}}$ and its distribution is completely characterized by that of $\mathbf{N}$. Let $\boldsymbol{\tau} \sim \pi_\tau$ and let $\pi_{\mathbf{N}}(\cdot|\boldsymbol{\tau})$ denote the conditional distribution of $\mathbf{N}$ given $\boldsymbol{\tau}$, so that the conditional distribution of $\mathbf{\Phi}$ given $\boldsymbol{\tau}$ is

$$\pi_{\Phi}(A|\boldsymbol{\tau}) = \pi_{\mathbf{N}}(n|\boldsymbol{\tau})\mathbf{1}_{\{A=\{1,\ldots,n\}\}}, \quad A \in \mathcal{F}_{\mathbb{N}} \tag{3.1}$$

Note that $|\mathbf{\Phi}| = \mathbf{N}$ and $\mathbf{1}_{\{j\in\mathbf{\Phi}\}} = \mathbf{1}_{\{j\leq\mathbf{N}\}}$, so (2.1) reduces to

$$\mathbf{w}_j = \mathbb{E}\left[\frac{1}{\mathbf{N}}\mathbf{1}_{\{j\leq\mathbf{N}\}}\,\middle|\,\boldsymbol{\tau}\right] = \sum_{n\geq j} \frac{\pi_{\mathbf{N}}(n|\boldsymbol{\tau})}{n}, \quad j \geq 1. \tag{3.2}$$

Clearly, $\mathbf{w}_i \geq \mathbf{w}_j$ almost surely, for every $i \leq j$. Thus, this type of distributions over $\mathcal{F}_{\mathbb{N}}$ will always derive in almost surely decreasing weights sequences. Distributions on the simplex based on random sets with no gaps have already been studied, for instance [7] showed that the choice $\boldsymbol{\tau} = \boldsymbol{\lambda}$ for some $\boldsymbol{\lambda} \sim \mathsf{Be}(\alpha, \beta)$, and

$$\pi_{\mathbf{N}}(n|\boldsymbol{\lambda}) = n\boldsymbol{\lambda}^2(1-\boldsymbol{\lambda})^{n-1}\mathbf{1}_{\{n\in\mathbb{N}\}} \tag{3.3}$$

gives

$$\mathbf{w}_j = \sum_{n \geq j} \boldsymbol{\lambda}^2(1-\boldsymbol{\lambda})^{n-1} = \boldsymbol{\lambda}(1-\boldsymbol{\lambda})^{j-1}, \quad j \geq 1, \tag{3.4}$$

so the decreasingly ordered Geometric weights are recovered. In [4] the authors analyse further examples such as $\boldsymbol{\lambda} \sim \mathsf{Ga}(a, b)$ and

$$\pi_{\mathrm{N}}(n|\boldsymbol{\lambda}) = \frac{e^{-\boldsymbol{\lambda}}\boldsymbol{\lambda}^{n-1}}{(n-1)!}\mathbf{1}_{\{n \in \mathbb{N}\}}, \tag{3.5}$$

in which case the weights simplify to

$$\mathbf{w}_j = \boldsymbol{\lambda}^{-1}e^{-\boldsymbol{\lambda}}\left(e^{\boldsymbol{\lambda}} - \sum_{k=0}^{j-1}\frac{\boldsymbol{\lambda}^k}{k!}\right). \tag{3.6}$$

In general, if π_{N} describes a parametric family whose support is $\mathbb{N}$, the conditions in Corollary 2.3 are satisfied, thus the corresponding prior has full support. As to the MCMC algorithm, we note that for random sets with no gaps, in order to update $\boldsymbol{\Phi}_k$, it suffices to sample $\mathbf{N}_k$ from

$$\pi(\mathbf{N}_k = n \mid \ldots) \propto \frac{\pi_{\mathrm{N}}(n \mid \boldsymbol{\tau})}{n}\mathbf{1}_{\{\mathbf{d}_k \leq n\}}, n \in \mathbb{N}, \tag{3.7}$$

and set $\boldsymbol{\Phi}_k = \{1, \ldots, \mathbf{N}_k\}$, this can be easily seen by embedding (3.1) into (2.7). Analogously, equation (2.8) reduces to

$$\pi(\boldsymbol{\tau} \mid \ldots) \propto \pi_{\tau}(\boldsymbol{\tau}) \prod_{k=1}^{m} \pi_{\mathrm{N}}(\mathbf{N}_k \mid \boldsymbol{\tau}). \tag{3.8}$$

In particular, for the model described by Eq. (3.3), the above posterior distributions simplify to $\pi(\mathbf{N}_k = n \mid \ldots) \propto (1-\boldsymbol{\lambda})^n\mathbf{1}_{\{\mathbf{d}_k \leq n\}}$, and $\pi(\boldsymbol{\lambda} \mid \ldots) \propto (\boldsymbol{\lambda})^{\alpha+2m-1}(1-\boldsymbol{\lambda})^{\beta+\sum_k \mathbf{N}_k - m - 1}\mathbf{1}_{\{\boldsymbol{\lambda} \in [0,1]\}}$. From which is easy to sample, as the former describes a truncated Geometric distribution and the latter a Beta distribution. As to the model that corresponds to Eq. (3.5), the distributions (2.7) and (2.8) reduce to $\pi(\mathbf{N}_k = n \mid \ldots) \propto \boldsymbol{\lambda}^n(n!)^{-1}\mathbf{1}_{\{\mathbf{d}_k \leq n\}}$ and $\pi(\boldsymbol{\lambda} \mid \ldots) \propto e^{-\boldsymbol{\lambda}(b+m)}\boldsymbol{\lambda}^{a+\sum_k \mathbf{N}_k - m - 1}\mathbf{1}_{\{\boldsymbol{\lambda} \in [0,\infty)\}}$, which are a truncated Poisson and a Gamma distribution, respectively.

3.2 Random Sets with Bernoulli Chosen Elements

Consider $\boldsymbol{\tau} = (\boldsymbol{\lambda}, \boldsymbol{v}_1, \boldsymbol{v}_2, \ldots)$ a collection of independent random variables, with $\boldsymbol{\lambda} \sim \pi_\lambda$, and $\boldsymbol{v}_i \sim \mathsf{Be}(\alpha_i, \beta_i)$. As above, let $\mathbf{N}$ be a random variable taking values in $\mathbb{N}$, whose conditional mass probability function given $\boldsymbol{\lambda}$ is $\pi_{\mathrm{N}}(\cdot|\boldsymbol{\lambda})$. Independently, for $i \geq 1$, let $\{\mathbf{z}_i|\boldsymbol{v}_i\} \sim \mathsf{Ber}(\boldsymbol{v}_i)$. Define $\boldsymbol{\Phi}$ as the set containing the $\mathbf{N}$ smallest indexes i such that $\mathbf{z}_i = 1$. That is

$$j \in \boldsymbol{\Phi} \quad \Leftrightarrow \quad \mathbf{z}_j = 1 \quad \text{and} \quad \sum_{i<j} \mathbf{z}_i < \mathbf{N}. \tag{3.9}$$

For example, say that for some realization $\mathbf{N} = 3$, $\mathbf{z}_1 = 1$, $\mathbf{z}_2 = 1$, $\mathbf{z}_3 = 0$, $\mathbf{z}_4 = 0$ and $\mathbf{z}_5 = 1$, then we would have $\boldsymbol{\Phi} = \{1, 2, 5\}$. In general, for $\boldsymbol{\Phi}$ to be well defined we require the events $(\mathbf{z}_i = 1)_{i\geq 1}$ to occur infinitely often, this assures that for every $n \in \mathbb{N}$ there exist at least n indexes i such that $\mathbf{z}_i = 1$. As $\mathbf{z}_1, \mathbf{z}_2, \ldots$ inherit the mutual independence of $\boldsymbol{v}_1, \boldsymbol{v}_2, \ldots$, whenever

$$\sum_{i\geq 1} \frac{\alpha_i}{\alpha_i + \beta_i} = \infty, \tag{3.10}$$

we have that

$$\sum_{i\geq 1} \mathbb{P}[\mathbf{z}_i = 1] = \sum_{i\geq 1} \mathbb{E}[\mathbb{P}[\mathbf{z}_i = 1|\boldsymbol{v}_i]] = \sum_{i\geq 1} \mathbb{E}[\boldsymbol{v}_i] = \infty,$$

and by Borel-Cantelli lemma, we obtain

$$\mathbb{P}\left[\limsup_{i\to\infty}\{\mathbf{z}_i = 1\}\right] = 1,$$

that is, $\mathbf{z}_i = 1$, for infinitely many i's. In such case, $\boldsymbol{\Phi}$ is well defined and the conditional distribution of $\boldsymbol{\Phi}$ given $\boldsymbol{\tau}$ decomposes as

$$\pi_\Phi(A|\boldsymbol{\tau}) = \pi_{\mathrm{N}}(n|\boldsymbol{\lambda}) \left[\prod_{i\in A} \boldsymbol{v}_i\right] \left[\prod_{i\in A'} (1 - \boldsymbol{v}_i)\right] \mathbf{1}_{\{|A|=n\}}, \quad A \in \mathcal{F}_{\mathbb{N}} \tag{3.11}$$

where $A' = \{i < \max A : i \notin A\}$.

In many instances the explicit computation of the weights is not necessary. For example, as explained in Sect. 2.1, in a density estimation context it suffices to characterize the latent random sets. Despite, being able to compute the weights can be very useful when studying the theoretical properties of the model. For π_Φ as in

Eq. (3.11) and $\boldsymbol{\tau} = (\boldsymbol{\lambda}, \boldsymbol{v}_1, \boldsymbol{v}_2, \ldots)$ as above, by Eqs. (2.1) and (3.9) we find that the corresponding weights simplify to

$$\begin{aligned} \mathbf{w}_j &= \mathbb{E}\left[\frac{1}{\mathbf{N}}\mathbf{1}_{\{\mathbf{z}_j=1\}}\mathbf{1}_{\left\{\sum_{i<j}\mathbf{z}_i<\mathbf{N}\right\}}\,\middle|\,\boldsymbol{\tau}\right] \\ &= \mathbb{E}\left[\frac{1}{\mathbf{N}}\mathbf{1}_{\{\mathbf{Z}^{(j)}<\mathbf{N}\}}\,\middle|\,\boldsymbol{\tau}\right]\mathbb{E}[\mathbf{1}_{\{\mathbf{z}_j=1\}}\mid \boldsymbol{v}_j] \\ &= \mathbb{E}\left[\frac{1}{\mathbf{N}}\mathbf{1}_{\{\mathbf{Z}^{(j)}<\mathbf{N}\}}\,\middle|\,\boldsymbol{\tau}\right]\boldsymbol{v}_j, \end{aligned} \tag{3.12}$$

where $\mathbf{Z}^{(j)} = \sum_{i<j}\mathbf{z}_i$, for $j \geq 1$. Obtaining an analytical expression for the weights, clearly depends on the possibility to compute $\mathbb{E}\left[\frac{1}{\mathbf{N}}\mathbf{1}_{\{\mathbf{Z}^{(j)}<\mathbf{N}\}}\middle|\boldsymbol{\tau}\right]$. For instance, if we let $\pi_{\mathrm{N}}(\cdot|\boldsymbol{\lambda})$ be as in (3.3) for some $\boldsymbol{\lambda} \sim \mathsf{Be}(\alpha, \beta)$, by the tower property of conditional expectation and following (3.4) we obtain

$$\begin{aligned} \mathbb{E}\left[\frac{1}{\mathbf{N}}\mathbf{1}_{\{\mathbf{Z}^{(j)}<\mathbf{N}\}}\,\middle|\,\boldsymbol{\tau}\right] &= \mathbb{E}\left[\boldsymbol{\lambda}(1-\boldsymbol{\lambda})^{\mathbf{Z}^{(j)}}\,\middle|\,\boldsymbol{\tau}\right] \\ &= \boldsymbol{\lambda}\prod_{i=1}^{j-1}\mathbb{E}\left[(1-\boldsymbol{\lambda})^{\mathbf{z}_i}\mid \boldsymbol{\lambda}, \boldsymbol{v}_i\right] \\ &= \boldsymbol{\lambda}\prod_{i=1}^{j-1}\left[(1-\boldsymbol{\lambda})\boldsymbol{v}_i + (1-\boldsymbol{v}_i)\right]. \end{aligned} \tag{3.13}$$

Hence, embedding (3.13) into (3.12), we get

$$\mathbf{w}_j = \boldsymbol{v}_j\boldsymbol{\lambda}\prod_{i=1}^{j-1}(1-\boldsymbol{v}_i\boldsymbol{\lambda}), \quad j \geq 1, \tag{3.14}$$

with the convention that the empty product equals 1. Note that π_{N} as in (3.3) satisfies the conditions of Corollary 2.3, thus, the prior corresponding to the weights given by (3.14) has full support. This proves the following result

Proposition 3.1 *Let* $\boldsymbol{\lambda} \sim \mathsf{Be}(\alpha, \beta)$ *and independently for* $i \geq 1$ *consider* $\boldsymbol{v}_i \sim \mathsf{Be}(\alpha_i, \beta_i)$ *for some* $\alpha, \beta, \alpha_i, \beta_i > 0$ *such that* $\sum_{i\geq 1}\alpha_i/(\alpha_i+\beta_i) = \infty$. *Fix* $\boldsymbol{\tau} = (\boldsymbol{\lambda}, \boldsymbol{v}_1, \boldsymbol{v}_2, \ldots)$ *and consider a random set* $\{\boldsymbol{\Phi} \mid \boldsymbol{\tau}\} \sim \pi_{\Phi}(\cdot \mid \boldsymbol{\tau})$, *for* π_{Φ} *as in* (3.11), *where* π_{N} *is as in* (3.3). *Define* $\mathbf{W} = (\mathbf{w}_j)_{j\geq 1}$ *through* (2.1). *Then,*

(i) $\mathbf{W} = (\mathbf{w}_j)_{j\geq 1}$ *is a stick-breaking sequence featuring conditionally independent length variables* $(\mathbf{v}_i = \boldsymbol{v}_i\boldsymbol{\lambda})_{i\geq 1}$.
(ii) *For every* $\varepsilon > 0$, $\mathbb{P}\left[\max_{j\geq 1}\mathbf{w}_j < \varepsilon\right] > 0$. *That is, any the prior with weights* $\mathbf{W}$ *has full support.*

Further theoretical properties of the model described in Proposition 3.1 will be analysed in the following section. For now we explain how to sample from (2.7) and (2.8) for this example. In order to update $\mathbf{\Phi}_k$, **note that it can be defined as the first** $\mathbf{N}_k$ indexes i's such that $\mathbf{z}_{k,i} = 1$, where $\mathbf{N}_k$ is sampled from $\pi_{\mathrm{N}}\left(\cdot|\boldsymbol{\lambda}\right)$ as in (3.3), and independently for $i \geq 1$, $\mathbf{z}_{k,i} \sim \mathsf{Ber}\left(\boldsymbol{v}_i\right)$. That is, $\mathbf{\Phi}_k = A$ if and only if $\mathbf{N}_k = |A|$, $\mathbf{z}_{k,i} = 1$ for every $i \in A$, and $\mathbf{z}_{k,i} = 0$ for every $i \in A' = \{i \leq \max A : i \notin A\}$. Thus, to sample from (2.7), we can equivalently sample from $\pi\left(\mathbf{N}_k = n, \mathbf{z}_{k,i} = z_i, i \geq 1 \,\middle|\, \ldots\right)$ and later construct $\mathbf{\Phi}_k$ as aforementioned. To this aim we compute

$$
\begin{aligned}
&\pi\left(\mathbf{N}_k = n, \mathbf{z}_{k,i} = z_i, i \geq 1 \,\middle|\, \ldots\right) \\
&\propto \frac{1}{n}\mathbf{1}_{\left\{z_{\mathbf{d}_k}=1\right\}}\mathbf{1}_{\left\{\sum_{i=1}^{\mathbf{d}_k} z_i - 1 < n\right\}}{}^{n}\,(\boldsymbol{\lambda})^2\,(1-\boldsymbol{\lambda})^{n-1}\prod_{i\geq 1}(\boldsymbol{v}_i)^{z_i}\,(1-\boldsymbol{v}_i)^{1-z_i} \\
&\propto \left[\prod_{i=1}^{\mathbf{d}_k-1}(\boldsymbol{v}_i)^{z_i}\,(1-\boldsymbol{v}_i)^{1-z_i}\right]\left[(1-\boldsymbol{\lambda})^{n-1}\mathbf{1}_{\left\{\sum_{i=1}^{\mathbf{d}_k} z_i \leq n\right\}}\right]\left[\mathbf{1}_{\left\{z_{\mathbf{d}_k}=1\right\}}\right]\times \\
&\qquad\qquad \times\left[\prod_{i>\mathbf{d}_k}(\boldsymbol{v}_i)^{z_i}\,(1-\boldsymbol{v}_i)^{1-z_i}\right] \\
&\propto \left[\prod_{i=1}^{\mathbf{d}_k-1}(\boldsymbol{v}_i(1-\boldsymbol{\lambda}))^{z_i}\,(1-\boldsymbol{v}_i)^{1-z_i}\right]\left[\frac{(1-\boldsymbol{\lambda})^{n-1}}{(1-\boldsymbol{\lambda})^{\sum_{i=1}^{\mathbf{d}_k} z_i - 1}}\mathbf{1}_{\left\{\sum_{i=1}^{\mathbf{d}_k} z_i \leq n\right\}}\right]\left[\mathbf{1}_{\left\{z_{\mathbf{d}_k}=1\right\}}\right]\times \\
&\qquad\qquad \times\left[\prod_{i>\mathbf{d}_k}(\boldsymbol{v}_i)^{z_i}\,(1-\boldsymbol{v}_i)^{1-z_i}\right].
\end{aligned}
\tag{3.15}
$$

Sampling from the above distribution can be easily achieved by first sampling $\mathbf{z}_{k,i} \sim \mathsf{Ber}\left(\tilde{\boldsymbol{v}}_i\right)$, where $\tilde{\boldsymbol{v}}_i = \boldsymbol{v}_i(1-\boldsymbol{\lambda})/\{\boldsymbol{v}_i(1-\boldsymbol{\lambda}) + (1-\boldsymbol{v}_i)\}$, for $i \leq \mathbf{d}_k - 1$, and setting $\mathbf{z}_{k,\mathbf{d}_k} = 1$. Then conditionally given $\sum_{i=1}^{\mathbf{d}_k}\mathbf{z}_{k,i}$, sample $\mathbf{N}_l^{(k)}$ from a truncated $\mathsf{Geo}\left(\boldsymbol{\lambda}\right)$. Finally, conditionally given $\mathbf{N}_k$, sample $\mathbf{z}_{k,i} \sim \mathsf{Ber}\left(\boldsymbol{v}_i\right)$ for $i > \mathbf{d}_k$ until $\sum_i \mathbf{z}_{k,i} = \mathbf{N}_k$.

Now, to update the random parameter of the sets, since $\boldsymbol{\tau} = (\boldsymbol{\lambda}, \boldsymbol{v}_1, \boldsymbol{v}_2, \ldots)$ is a sequence of independent random variables, we first sample $\boldsymbol{\lambda}$ from

$$
\begin{aligned}
\pi\left(\boldsymbol{\lambda} \mid \ldots\right) &\propto \mathsf{Be}\left(\boldsymbol{\lambda}\middle|\alpha,\beta\right)\prod_{k=1}^{m}\pi_{\mathrm{N}}\left(\mathbf{N}_k \mid \boldsymbol{\lambda}\right) \\
&\propto (\boldsymbol{\lambda})^{\alpha+2m-1}\,(1-\boldsymbol{\lambda})^{\beta+\sum_{k=1}^{m}\mathbf{N}_k - m - 1}
\end{aligned}
$$

which is a $\mathsf{Be}\left(\alpha + 2m, \beta + \sum_{k=1}^{m} \mathbf{N}_k - m\right)$ distribution, and latter, independently for $i \geq 1$, we sample $\boldsymbol{v}_i$ from

$$\begin{aligned}
\pi\left(\boldsymbol{v}_i \mid \ldots\right) &\propto \mathsf{Be}\left(\boldsymbol{v}_i \mid \alpha_i, \beta_i\right) \prod_{k=1}^{m} \pi_\Phi\left(\boldsymbol{\Phi}_k \mid \boldsymbol{\tau}\right) \\
&\propto \mathsf{Be}\left(\boldsymbol{v}_i \mid \alpha_i, \beta_i\right) \prod_{k\in\varphi_i} \left(\boldsymbol{v}_i\right)^{\mathbf{z}_{k,i}} \left(1 - \boldsymbol{v}_i\right)^{1-\mathbf{z}_{k,i}} \\
&\propto \left(\boldsymbol{v}_i\right)^{\alpha_i + \sum_{k\in\varphi_i} \mathbf{z}_{k,i} - 1} \left(1 - \boldsymbol{v}_i\right)^{\beta_i + |\varphi_i| - \sum_{k\in\varphi_i} \mathbf{z}_{k,i} - 1},
\end{aligned}$$

that describes a $\mathsf{Be}\left(\alpha_i + \sum_{k\in\varphi_i} \mathbf{z}_{k,i}, \; \beta_i + |\varphi_i| - \sum_{k\in\varphi_i} \mathbf{z}_{k,i}\right)$, where $\varphi_i = \{k : i \leq \max \boldsymbol{\Phi}_k\}$.

4 Geometric Product Stick-Breaking Processes

To any weight sequence $\mathbf{W} = (\mathbf{w}_j)_{j\geq 1}$ as in (3.14) for some independent random variables $\boldsymbol{\lambda} \sim \mathsf{Be}(\alpha, \beta)$ and $\boldsymbol{v}_i \sim \mathsf{Be}(\alpha_i, \beta_i)$, such that (3.10) holds we call a *Geometric product stick-breaking weights sequence* (GPSBW). To any random mixture, $\mathbf{p}$, as in (1.1) whose weights sequence is a GPSBW, we call a *Geometric product stick-breaking mixture* (GPSBM). In particular if $\psi(\cdot|s) = \delta_s$ we call $\mathbf{p}$ a *Geometric product stick-breaking process* (GPSBP). The first thing to mention is that by construction, equation (2.2), and Proposition 3.1 we have the following result.

Corollary 4.1 *The elements of any* GPSBW *sum up to* 1, *and any* GPSBP *has full support.*

Notice that for $\boldsymbol{\lambda}$ and $(\boldsymbol{v}_i)_{i\geq 1}$ as above,

$$\hat{\mathbf{w}}_j = \boldsymbol{\lambda}(1 - \boldsymbol{\lambda})^{j-1}, \quad j \geq 1 \tag{4.1}$$

define the Geometric weights, whilst

$$\tilde{\mathbf{w}}_j = \boldsymbol{v}_j \prod_{i=1}^{j-1} (1 - \boldsymbol{v}_i), \quad j \geq 1 \tag{4.2}$$

form a stick-breaking sequence featuring independent length variables. So each length variable of GPSBW is precisely the product of a length variable of $\hat{\mathbf{W}} = (\hat{\mathbf{w}}_j)_{j\geq 1}$ and one of $\tilde{\mathbf{W}} = (\tilde{\mathbf{w}}_j)_{j\geq 1}$. Recalling that a $\mathsf{Be}(a, b)$ distribution converges weakly to δ_1, as $a \to \infty$ or $b \to 0$, it is clear that $\mathbf{v}_i = \boldsymbol{\lambda}\boldsymbol{v}_i$ converges in distribution to $\boldsymbol{\lambda}$ or to $\boldsymbol{v}_i$ for extreme values of the parameters α, β, $(\alpha_i, \beta_i)_{i\geq 1}$. This convergence property is inherited to the weights sequences, for example, if $\alpha_i = \alpha = 1$, and $\beta_i = \theta > 0$, as $\beta \to 0$ the GPSBW's converges in distribution to the Dirichlet weights. If otherwise we fix $\beta > 0$, as $\theta \to 0$ the GPSBW's converge in distribution to the Geometric weights. Although this is a consequence of elementary properties of the Beta distribution, for the sake of completeness we include the proof below.

Proposition 4.2 *For every* $i \geq 1$ *let* $\mathbf{v}_i^{(\alpha,\beta,\alpha_i,\beta_i)} = \boldsymbol{\lambda}^{(\alpha,\beta)}\boldsymbol{v}_i^{(\alpha_i,\beta_i)}$ *for some independent random variables* $\left(\boldsymbol{v}_i^{(\alpha_i,\beta_i)} \sim \mathsf{Be}(\alpha_i,\beta_i)\right)_{i\geq 1}$ *and* $\boldsymbol{\lambda}^{(\alpha,\beta)} \sim \mathsf{Be}(\alpha,\beta)$. *Then*

(a) *For every* $(\alpha_i,\beta_i)_{i\geq 1}$ *and* α *fixed,* $\left(\mathbf{v}_i^{(\alpha,\beta,\alpha_i,\beta_i)}\right)_{i\geq 1} \stackrel{d}{\to} \left(\boldsymbol{v}_i^{(\alpha_i,\beta_i)}\right)_{i\geq 1}$, *as* $\beta \to 0$.

(b) *For every* $(\alpha_i,\beta_i)_{i\geq 1}$ *and* β *fixed,* $\left(\mathbf{v}_i^{(\alpha,\beta,\alpha_i,\beta_i)}\right)_{i\geq 1} \stackrel{d}{\to} \left(\boldsymbol{v}_i^{(\alpha_i,\beta_i)}\right)_{i\geq 1}$, *as* $\alpha \to \infty$.

(c) *For every* $(\alpha_i)_{i\geq 1}$, α *and* β *fixed,* $\left(\mathbf{v}_i^{(\alpha,\beta,\alpha_i,\beta_i)}\right)_{i\geq 1} \stackrel{d}{\to} \left(\boldsymbol{\lambda}^{(\alpha,\beta)},\boldsymbol{\lambda}^{(\alpha,\beta)},\ldots\right)$, *as* $\max_i \beta_i \to 0$.

(d) *For every* $(\beta_i)_{i\geq 1}$, α *and* β *fixed,* $\left(\mathbf{v}_i^{(\alpha,\beta,\alpha_i,\beta_i)}\right)_{i\geq 1} \stackrel{d}{\to} \left(\boldsymbol{\lambda}^{(\alpha,\beta)},\boldsymbol{\lambda}^{(\alpha,\beta)},\ldots\right)$, *as* $\min_i \alpha_i \to \infty$.

Proof Fix $(\alpha_i,\beta_i)_{i\geq 1}$. Let $(t_1,\ldots,t_n) \in \mathbb{R}^n$, and let us denote $\Sigma(n) = \sum_{i=1}^n t_i \boldsymbol{v}_i^{(\alpha_i,\beta_i)}$. Then

$$\begin{aligned}
\mathbb{E}\left[\exp\left\{\sum_{i=1}^n t_i \mathbf{v}_i^{(\alpha,\beta,\alpha_i,\beta_i)}\right\}\right] &= \mathbb{E}\left[\exp\left\{\boldsymbol{\lambda}^{(\alpha,\beta)}\sum_{i=1}^n t_i \boldsymbol{v}_i^{(\alpha_i,\beta_i)}\right\}\right] \\
&= \mathbb{E}\left[\mathbb{E}\left[\exp\left\{\boldsymbol{\lambda}^{(\alpha,\beta)}\Sigma(n)\right\}\middle|\Sigma(n)\right]\right] \\
&= \mathbb{E}\left[1+\sum_{k=1}^\infty\left(\prod_{r=0}^{k-1}\frac{\alpha+r}{\alpha+\beta+r}\right)\frac{\Sigma(n)^k}{k!}\right].
\end{aligned}$$

As $e^{\Sigma(n)} \leq \prod_{i=1}^n e^{t_i} < \infty$ a.s., by Lebesgue dominated convergence theorem we obtain that for fixed α, as $\beta \to 0$

$$\mathbb{E}\left[\exp\left\{\sum_{i=1}^n t_i \mathbf{v}_i^{(\alpha,\beta,\alpha_i,\beta_i)}\right\}\right] \to \mathbb{E}\left[e^{\Sigma(n)}\right] = \mathbb{E}\left[\exp\left\{\sum_{i=1}^n t_i \boldsymbol{v}_i^{(\alpha_i,\beta_i)}\right\}\right],$$

that is $\left(\mathbf{v}_i^{(\alpha,\beta,\alpha_i,\beta_i)}\right)_{i=1}^n \stackrel{d}{\to} \left(\boldsymbol{v}_i^{(\alpha_i,\beta_i)}\right)_{i=1}^n$, and same holds for β fixed, as $\alpha \to \infty$. Since n was arbitrary, we have proven (a) and (b).

To prove (c) and (d) fix $\alpha, \beta > 0$. Let $(t_1, \ldots, t_n) \in \mathbb{R}^n$, then

$$\begin{aligned}\mathbb{E}\left[\exp\left\{\sum_{i=1}^{n} t_i \mathbf{v}_i^{(\alpha,\beta,\alpha_i,\beta_i)}\right\}\right] &= \mathbb{E}\left[\mathbb{E}\left[\exp\left\{\sum_{i=1}^{n} \boldsymbol{\lambda}^{(\alpha,\beta)} t_i \boldsymbol{v}_i^{(\alpha_i,\beta_i)}\right\} \Big| \boldsymbol{\lambda}^{(\alpha,\beta)}\right]\right]\\ &= \mathbb{E}\left[\prod_{i=1}^{n} \mathbb{E}\left[\exp\left\{\boldsymbol{\lambda}^{(\alpha,\beta)} t_i \boldsymbol{v}_i^{(\alpha_i,\beta_i)}\right\} \Big| \boldsymbol{\lambda}^{(\alpha,\beta)}\right]\right]\\ &= \mathbb{E}\left[\prod_{i=1}^{n}\left\{1 + \sum_{k=1}^{\infty}\left(\prod_{r=0}^{k-1}\frac{\alpha_i + r}{\alpha_i + \beta_i + r}\right)\frac{(\boldsymbol{\lambda}^{(\alpha,\beta)} t_i)^k}{k!}\right\}\right].\end{aligned}$$

As $\prod_{i=1}^{n} e^{\boldsymbol{\lambda} t_i} \leq \prod_{i=1}^{n} e^{t_i} < \infty$ a.s. and for fixed i, $e^{\boldsymbol{\lambda} t_i} \leq e^{t_i} < \infty$ a.s., by Lebesgue dominated convergence theorem, for $(\alpha_i)_{i\geq 1}$ fixed, as $\max_i \beta_i \to 0$, we obtain

$$\mathbb{E}\left[\exp\left\{\sum_{i=1}^{n} t_i \mathbf{v}_i^{(\alpha,\beta,\alpha_i,\beta_i)}\right\}\right] \to \mathbb{E}\left[\prod_{i=1}^{n} e^{\boldsymbol{\lambda}^{(\alpha,\beta)} t_i}\right] = \mathbb{E}\left[\exp\left\{\sum_{i=1}^{n} \boldsymbol{\lambda}^{(\alpha,\beta)} t_i\right\}\right],$$

that is $\left(\mathbf{v}_i^{(\alpha,\beta,\alpha_i,\beta_i)}\right)_{i=1}^{n} \xrightarrow{d} \left(\boldsymbol{\lambda}^{(\alpha,\beta)}\right)_{i=1}^{n}$, and the same holds if we fix $(\beta_i)_{i\geq 1}$ and let $\min_i \alpha_i \to \infty$. Since n was arbitrary, this proves (c) and (d). □

Remark 4.3 Insomuch as the mapping

$$(v_1, v_2 \ldots, v_j) \mapsto \left(v_1, v_2(1 - v_1), \ldots, v_j \prod_{i=1}^{j-1}(1 - v_i)\right)$$

is continuous with respect to the product topology, and under mild conditions of ψ,

$$\{(w_1, w_2, \ldots), (s_1, s_2, \ldots)\} \mapsto \sum_{j\geq 1} w_j \psi(\cdot|s_j),$$

is continuous with respect to the topology of weak convergence (see Appendix A), similar limiting properties to that of Proposition 4.2 are inherited to the corresponding GPSBW's and GPSBM's.

For extreme values of the Beta parameters, $\mathbf{W}$ as in (3.14) converges in distribution to $\hat{\mathbf{W}}$ as in (4.1) and $\tilde{\mathbf{W}}$ as in (4.2), this does not mean GPSBW's are some kind of hybrid between Geometric weights sequences and stick-breaking sequences with independent length variables. In fact, GPSBW's are less likely to be decreasingly ordered, and have heavier tails than their limiting counterparts. Explicitly, for each every $j \geq 1$, the event

$$(\mathbf{w}_j > \mathbf{w}_{j+1}) = (\boldsymbol{v}_j > \boldsymbol{v}_{j+1} - \boldsymbol{\lambda}\boldsymbol{v}_j\boldsymbol{v}_{j+1}) \subseteq (\boldsymbol{v}_j > \boldsymbol{v}_{j+1} - \boldsymbol{v}_j\boldsymbol{v}_{j+1}) = (\tilde{\mathbf{w}}_j > \tilde{\mathbf{w}}_{j+1}),$$

hence $\mathbb{P}[\mathbf{w}_j > \mathbf{w}_{j+1}] \leq \mathbb{P}[\tilde{\mathbf{w}}_j > \tilde{\mathbf{w}}_{j+1}]$, and trivially, $\mathbb{P}[\mathbf{w}_j > \mathbf{w}_{j+1}] \leq 1 = \mathbb{P}[\hat{\mathbf{w}}_j > \hat{\mathbf{w}}_{j+1}]$ holds as well. As to the tails of these random discrete distributions, we have that $(1 - \boldsymbol{\lambda v}_i) > (1 - \boldsymbol{v}_i)$ almost surely for every $i \geq 1$, thus

$$\sum_{j \geq n} \mathbf{w}_j = \prod_{j \geq n-1} (1 - \boldsymbol{\lambda v}_j) > \prod_{j \geq n-1} (1 - \boldsymbol{v}_j) = \sum_{j \geq n} \tilde{\mathbf{w}}_j$$

almost surely for every $n \geq 2$, and analogously $\sum_{j \geq n} \mathbf{w}_j > \sum_{j \geq n} \hat{\mathbf{w}}_j$. To illustrate this, in the right side of Fig. 1 we show three independent simulations of $(\mathbf{w}_j)_{j=1}^{20}$ (dashed lines), $(\tilde{\mathbf{w}}_j)_{j=1}^{20}$ (dot-dashed lines) and $(\hat{\mathbf{w}}_j)_{j=1}^{20}$ (solid lines), with parameters $\alpha = \alpha_i = 1$ and $\beta = \beta_i = 5$, for every $i \geq 1$. Inhere, we can descry

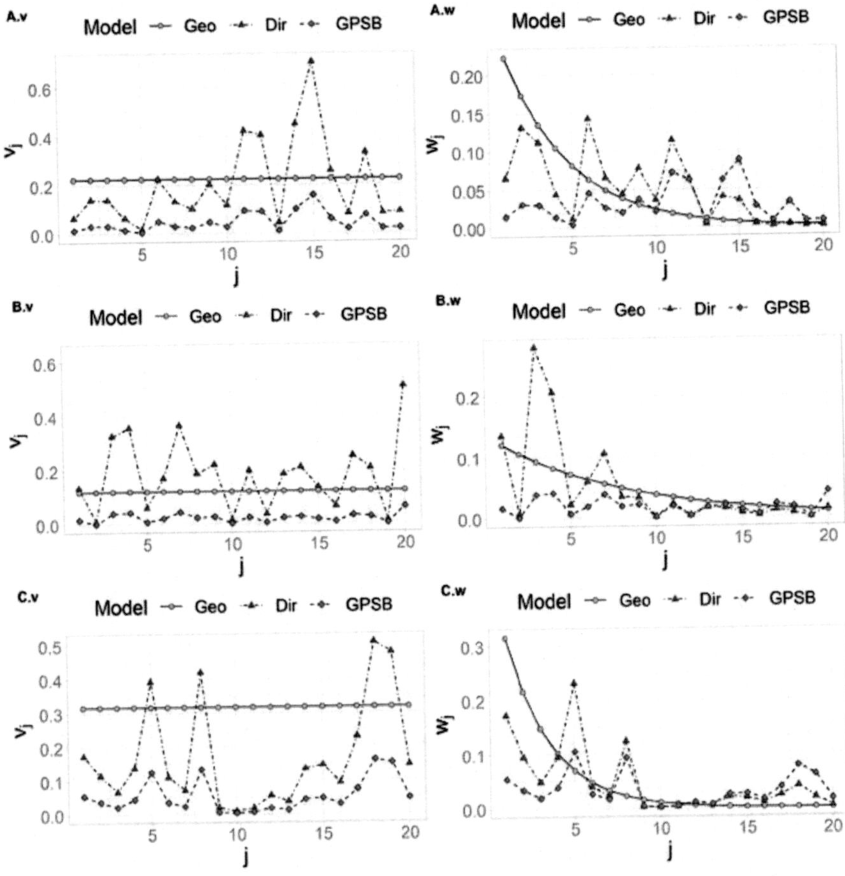

Fig. 1 In the left side (A.v, B.v, and C.v) we exhibit three independent simulations of the length variables $(\boldsymbol{\lambda v}_i)_{i=1}^{20}$ (dashed lines), $(\boldsymbol{v}_i)_{i=1}^{20}$ (dot-dashed lines) and $(\boldsymbol{\lambda})_{i=1}^{20}$ (solid lines), where the parameters were fixed to $\alpha = \alpha_i = 1$ and $\beta = \beta_i = 5$, for every $i \geq 1$, in all cases. The graphs in the right side (A.w, B.w, and C.w) show the corresponding stick-breaking weights, respectively

that $\hat{\mathbf{w}}_j \geq \hat{\mathbf{w}}_{j+1}$ for every $j \geq 1$, and that whenever $\mathbf{w}_j \geq \mathbf{w}_{j+1}$ we also have that $\hat{\mathbf{w}}_j \geq \hat{\mathbf{w}}_{j+1}$, but the converse is not true, for in example, in A.w, $\hat{\mathbf{w}}_{14} \geq \hat{\mathbf{w}}_{15}$ and $\mathbf{w}_{14} \leq \mathbf{w}_{15}$. In the same figure it can be appreciated that $\sum_{j=1}^{20} \mathbf{w}_j < \sum_{j=1}^{20} \hat{\mathbf{w}}_j$ and $\sum_{j=1}^{20} \mathbf{w}_j < \sum_{j=1}^{20} \tilde{\mathbf{w}}_j$, since each weights sequence sums up to 1, this must mean that $(\mathbf{w}_j)_{j\geq 1}$ accumulates more mass after the index 20 than $(\hat{\mathbf{w}}_j)_{j\geq 1}$ and $(\tilde{\mathbf{w}}_j)_{j\geq 1}$ do.

4.1 Illustrations

To test the performance of GPSBM's in the context of Sect. 2.1, we simulated a dataset containing 220 observations, and estimate its density through five distinct mixtures. In all cases we assume a Gaussian kernel with random location and scale parameters, so that for each $j \geq 1$, $\boldsymbol{\xi}_j = (\mathbf{m}_j, \mathbf{q}_j)$, and $\psi(\cdot \mid \boldsymbol{\xi}_j) = \mathsf{N}(\mathbf{m}_j, \mathbf{q}_j^{-1})$. To attain a conjugacy for f and p_0 as in (2.5), we assume $p_0(\boldsymbol{\xi}_j) = \mathsf{N}(\mathbf{m}_j|\vartheta, \tau\mathbf{q}_j^{-1})\mathsf{Ga}(\mathbf{q}_j|a, b)$, where ϑ is the mean of the data, $\tau = 100$ and $a = b = 0.5$. Two of the adjusted mixtures correspond to a Geometric mixture with length variable $\boldsymbol{\lambda} \sim \mathsf{Be}(1, 1)$, and a Dirichlet mixture with parameter $\theta = 1$, that is, a stick-breaking mixtures with i.i.d. length variables $(\boldsymbol{v}_i)_{i\geq 1} \overset{\text{iid}}{\sim} \mathsf{Be}(1, 1)$. The remaining models are GPSBM's where $\boldsymbol{\lambda} \sim \mathsf{Be}(1, \beta)$ and $\boldsymbol{v}_i \sim \mathsf{Be}(1, \theta)$ independently for $i \geq 1$. For one of these mixtures we fixed $(\beta, \theta) = (1, 1)$ for a second one we took $(\beta, \theta) = (0.5, 1)$, and for the last one $(\beta, \theta) = (1, 0.5)$. For the sake of simplicity, and taking advantage of Proposition 4.2 and Remark 4.3, in Fig. 2 we also refer to the Geometric and Dirichlet mixtures as GPSBM's with $(\beta, \theta) = (1, 0)$ and $(\beta, \theta) = (0, 1)$, respectively.

In Fig. 2 we observe that the Dirichlet mixture (dashed lines) confuses two of the modes featured in the histogram of the data, the same happens with the GPSBM with $\beta = 0.5$ (long dash-dotted lines) which a priori behaves similar to the Dirichlet model. It is well-known that this type of problems arise when the total mass parameter, θ, of the Dirichlet process is not carefully choosen for a dataset. The three remaining models correct this issue, and overall seem adapt well enough to the data. Naively, the GPSBM with parameters $\beta = \theta = 1$, appears to be the one that performs better, given that at the high-density areas, the Geometric mixture (solid line) and the GPSBM with parameter $\theta = 0.5$ (long dashed line) differ slightly from the rest of the models. If we focus in the Dirichlet mixture, the Geometric mixture and the GPSBM with $\beta = \theta = 1$, it should be expected that the latter performs better than its limiting counterparts. This, due to the fact that a priori the GPSBW has a heavier tail, leading to a more flexible mixture. However, we must mention that this flexibility comes with a cost at efficiency, in the sense that the GPSBM is prone to use more mixture components to estimate the density than the Geometric and Dirichlet mixtures.

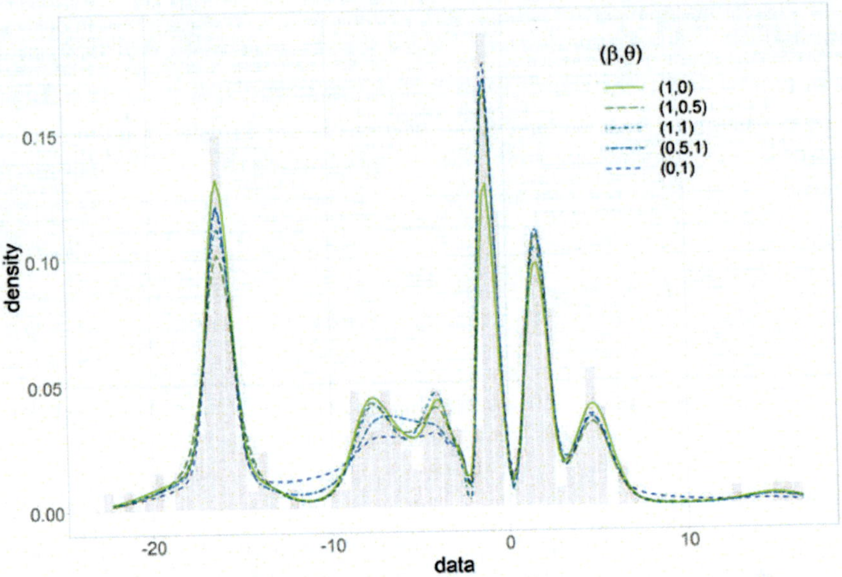

Fig. 2 Histogram of the data, and estimated densities after 5000 iterations of the Gibbs sampler, with a burn-in period of 2000, for a Geometric mixture ($(\beta, \theta) = (1, 0)$), a Dirichlet mixture ($(\beta, \theta) = (0, 1)$) and three GPSBM's

5 Final Remarks

We took a relatively simple idea, originally used for simulation purposes, and exploit it to define a novel class of Bayesian non-parametric priors. Taking advantage of the context in which the construction arose, we propose an estimation method via Gibbs sampler implementation.

In contrast to other models that rely on a latent random sets with no gaps, and that have already been analysed in the literature. The general class considered, not only recovers such random probability measures, but can also lead to sequences of weights with heavier tails. From a Bayesian non-parametric perspective there are drawbacks of working with heavy tailed weights, for example if truncation is necessary, working with weights that decrease slower makes the truncation error decrease slower as well. Another disadvantage is that for density estimation purposes, and when the data features few modes, models with slowly decreasing weights tend to use an unnecessarily large amount of mixture components, so in this sense the model losses efficiency. On the other side, a nice advantage of heavily tailed weights is that the estimated density will very likely recognize every mode in the data, while other models might not, so for example if we encounter a dataset whose histogram features a large amount of modes, it seems like a good idea to use a mixture model with heavily tailed weights.

Now, in essence, the random set construction builds a distribution over the infinite dimensional simplex, inhere we used this weights' distribution to define the prior of a random mixture. However, we shall highlight that distributions on the infinite dimensional simplex, can be used in a wide variety of contexts. For example, a distribution, π_{W}, over Δ_∞ automatically determines a the law of a random partition

$$\mathbf{\Pi} = \{[0, \mathbf{w}_1), [\mathbf{w}_1, \mathbf{w}_1 + \mathbf{w}_2), [\mathbf{w}_1 + \mathbf{w}_2, \mathbf{w}_1 + \mathbf{w}_2 + \mathbf{w}_3), \ldots\},$$

of the interval $[0, 1)$, where $(\mathbf{w}_j)_{j\geq 1} \sim \pi_{\mathrm{W}}$, which can be useful in theoretical and applied stochastic models. In this context heavy-tailed weights would allow the latter blocks of $\mathbf{\Pi}$ to have bigger sizes. This said, while working with slowly deceasing weights in a Bayesian non-parametric context can have significant drawbacks, we shall highlight that in other applications it can be extremely interesting.

Acknowledgments We thank an anonymous referee for his/her careful review that led to substantial improvements, and to the project PAPIIT-UNAM IG100221. Many thanks to the CONACyT PhD scholarship program and the support of CONTEX project 2018-9B.

Appendix

The purpose of this section is to prove that under mild conditions of ψ, the mapping

$$\{(w_1, w_2, \ldots), (s_1, s_2, \ldots)\} \mapsto \sum_{j\geq 1} w_j \psi(\cdot|s_j),$$

is continuous with respect to the weak topology. This assures that if $\mathbf{W}^{(n)} = \left(\mathbf{w}_j^{(n)}\right)_{j\geq 1}$ converges in distribution to $\mathbf{W} = \left(\mathbf{w}_j\right)_{j\geq 1}$, and $\Xi^{(n)} = \left(\xi_j^{(n)}\right)_{j\geq 1}$ converges in distribution to $\Xi = \left(\xi_j\right)_{j\geq 1}$, then $\sum_{j\geq 1} \mathbf{w}_j^{(n)} \psi\left(\cdot \,\middle|\, \xi_j^{(n)}\right)$ converges weakly in distribution to $\sum_{j\geq 1} \mathbf{w}_j \,\psi(\cdot \mid \xi_j)$.

Proposition A.1 *Let $(\mathcal{S}, \mathcal{B}(\mathcal{S}))$ and $(\mathcal{R}, \mathcal{B}(\mathcal{R}))$ be Polish spaces and let ψ be a probability kernel from $\mathcal{S}$ into $\mathcal{R}$ such that $\psi(\cdot \mid s_n)$ converges weakly to $\psi(\cdot \mid s)$ as $s_n \to s$ in $\mathcal{S}$. Then the mapping*

$$\{(w_1, w_2, \ldots), (s_1, s_2, \ldots)\} \mapsto \sum_{j\geq 1} w_j \psi(\cdot|s_j),$$

from $\Delta_\infty \times \mathcal{S}^\infty$ into the space of all probability measures over $(\mathcal{S}, \mathcal{B}(\mathcal{S}))$ is continuous with respect to the weak topology.

Proof Let $W = (w_1, w_2, \ldots)$, $\left\{W^{(n)} = \left(w_1^{(n)}, w_2^{(n)}, \ldots\right)\right\}_{n\geq 1}$ be elements of Δ_∞, and $S = (s_1, s_2, \ldots)$, $\left\{S^{(n)} = \left(s_1^{(n)}, s_2^{(n)}, \ldots\right)_{n\geq 1}\right\}$, be elements of $\mathcal{S}^\infty$, such that $w_j^{(n)} \to w_j$ and $s_j^{(n)} \to s_j$, for every $j \geq 1$. Define $p^{(n)} = \sum_{j\geq 1} w_j^{(n)} \psi\left(\cdot \middle| s_j^{(n)}\right)$ and $p = \sum_{j\geq 1} w_j \, \psi(\cdot \mid s_j)$. By the Portmanteau theorem (see for instance [1],[12] or [15]) it suffices to prove that for every continuous and bounded function $f : \mathcal{S} \to \mathbb{R}$,

$$p^{(n)}(f) = \int f(s)\, p^{(n)}(ds) \to \int f(s)\, dp(s) = p(f).$$

So fix a continuous and bounded function $f : S \to \mathbb{R}$. First note that by hypothesis $\psi\left(\cdot \middle| s_j^{(n)}\right)$ converges weakly to $\psi(\cdot \mid s_j)$, for every $j \geq 1$, by the Portmanteau theorem this implies

$$\psi\left(f \middle| s_j^{(n)}\right) = \int f(s)\, \psi\left(ds \middle| s_j^{(n)}\right) \to \int f(s)\, \psi(ds \mid s_j) = \psi(f \mid s_j),$$

thus $w_j^{(n)} \psi\left(f \middle| s_j^{(n)}\right) \to w_j \psi(f \mid s_j)$, for $j \geq 1$. Since f is bounded, there exist M such that $|f| \leq M$, hence, using the fact that $\psi\left(\cdot \middle| s_j^{(n)}\right)$ is a probability measure, we obtain

$$\left| w_j^{(n)} \psi\left(f \middle| s_j^{(n)}\right)\right| \leq w_j^{(n)} \psi\left(|f| \middle| s_j^{(n)}\right) \leq w_j^{(n)} M,$$

for every $n, j \geq 1$. Evidently, $M w_j^{(n)} \to M w_j$, and $\sum_{j\geq 1} M w_j^{(n)} = M = \sum_{j\geq 1} M w_j$. Then, by general Lebesgue dominated convergence theorem, we conclude

$$p^{(n)}(f) = \sum_{j\geq 1} w_j^{(n)} \psi\left(f \middle| s_j^{(n)}\right) \to \sum_{j\geq 1} w_j \psi(f \mid s_j) = p(f).$$

□

References

1. Billingsley, P.: Convergence of Probability Measures. Wiley Series in Probability And Statistics. John Wiley and Sons Inc., New York (1968)
2. Bissiri, P.G., Ongaro, A.: On the topological support of species sampling priors. Electron. J. Stat. **8**, 861–882 (2014)
3. Blackwell, D., MacQueen, J.: Ferguson distributions via Pólya urn schemes. Ann. Stat. **1**, 353–355 (1973)

4. De Blasi, P., Martínez, A.F., Mena, R.H., Prünster, I.: On the inferential implications of decreasing weight structures in mixture models. Comput. Stat. Data Anal. **147**, 106940 (2020)
5. Ferguson, T.S.: A Bayesian analysis of some nonparametric problems. Ann. Stat. **1**, 209–230 (1973)
6. Frühwirth-Schnatter, S., Celeux, G., Robert, C.P.: Handbook of Mixture Analysis. Chapman and Hall/CRC, Boca Raton (2019)
7. Fuentes-García, R., Mena, R.H., Walker, S.G.: A new Bayesian Nonparametric mixture model. Commun. Stat. Simul. Comput. **39**, 669–682 (2010)
8. Ghosal, S., van der Vaart, A.: Fundamentals of Nonparametric Bayesian Inference. Cambridge Series in Statistical and Probabilistic Mathematics. Cambridge University Press, Cambridge (2017)
9. Hjort, N., Holmes, C., Müller, P., Walker, S.G.: Bayesian Nonparametrics. Cambridge Series in Statistical and Probabilistic Mathematics. Cambridge University Press, Cambridge (2010)
10. Ishwaran, H., James, L.F.: Gibbs sampling methods for stick-breaking priors. J. Am. Stat. Assoc. **96**, 161–173 (2001)
11. James, L.F., Lijoi, A., Prünster, I.: Posterior analysis for normalized random measures with independent increments. J. R. Stat. Soc. B (Stat. Methodol.) **36**, 76–97 (2009)
12. Kallenberg, O.: Random Measures, Theory and Applications. Probability Theory and Stochastic Modelling, vol. 77, pp. 1–680. Springer International Publishing, Cham (2017)
13. Kingman, J.F.C.: Random discrete distributions (with discussion). J. R. Stat. Soc. B (Stat. Methodol.) **37**, 1–22 (1975)
14. McCloskey, T.: A model for the distribution of individuals by species in an environment. Michigan State University Department of Statistics (1965)
15. Parthasarathy, K.R.: Probability Measures on Metric Spaces. Academic Press, New York (1967)
16. Pitman, J.: Some developments of the Blackwell-MacQueen urn scheme. In: Ferguson, T.S., Shapley, L.S., MacQueen, J.B. (eds.) Statistics Probability and Game Theory. IMS Lecture Notes Monograph Series, vol. 30, pp. 245–267 (1996)
17. Pitman, J.: Combinatorial stochastic processes. In: École d'été de probabilités de Saint-Flour, vol. 1875, pp. 1–260. Springer, Berlin Heidelberg (2006)
18. Regazzini, E., Lijoi, A., Prünster, I.: Distributional results for means of normalized random measures with independent increments. Ann. Stat. **31**, 560–585 (2003)
19. Sethuraman, J.: A constructive definition of Dirichlet priors. Stat. Sin. **4**, 639–650 (1994)
20. Walker, S.G.: Sampling the Dirichlet mixture model with slices. Commun. Stat. Simul. Comput. **36**, 45–54 (2007)

Conjugate Predictive Distributions and Generalized Entropies

Eduardo Gutiérrez-Peña and Manuel Mendoza

Abstract It is well-known that maximizing the Shannon entropy gives rise to an exponential family of distributions. On the other hand, some Bayesian predictive distributions, derived from exponential family sampling models with standard conjugate priors on the canonical parameter, maximize a generalized entropy indexed by a parameter α. As $\alpha \to \infty$, this generalized entropy converges to the usual Shannon entropy, while the predictive distribution converges to its corresponding sampling model. The aim of this paper is to study this type of connection between generalized entropies based on a certain family of α-divergences and the class of predictive distributions mentioned above. We discuss two important examples in some detail, and argue that similar results must also hold for other exponential families.

Keywords Conjugate distribution · Deformed exponential function · Deformed logarithm · Exponential family · Shannon entropy

1 Introduction

The maximum entropy principle is often called up in situations where one wishes to select a single distribution P_* for a random variable X, as a representative of a class Γ of probability distributions, with the fewest assumptions about the true distribution of X. This is often done for purposes of inductive inference or for choosing an optimal decision; see, for example, Jaynes [13]. As discussed by Grünwald and Dawid [9], the distribution P_* that maximizes the entropy over Γ

E. Gutiérrez-Peña (✉)
Departamento de Probabilidad y Estadística, Universidad Nacional Autónoma de México, Mexico City, Mexico
e-mail: eduardo@sigma.iimas.unam.mx

M. Mendoza
Departamento de Estadística, Instituto Tecnológico Autónomo de México, Mexico City, Mexico
e-mail: mendoza@itam.mx

D. Hernández-Hernández et al. (eds.), *Advances in Probability and Mathematical Statistics*, Progress in Probability 79, https://doi.org/10.1007/978-3-030-85325-9_6

also minimizes the worst-case expected logarithmic score. In the terminology of statistical decision theory, P_* is a robust Bayes (or Γ-minimax) action when loss is measured by the logarithmic score. This provides a decision-theoretic interpretation of maximum entropy.

The choice of the class Γ is typically based on partial prior information available to the analyst. This information must be accounted for, but otherwise the selected distribution is sought to be as non-informative as possible. Partial information is usually described in terms of constraints on (some of) the moments of the distribution. Here we assume that X is defined on $\mathbb{R}$, and take $\Gamma = \Gamma_d \equiv \{P : E_P(S) = \varsigma\}$ where $\varsigma \in \mathbb{R}^d$ and $S = s(X)$ is a statistic taking values on $\mathbb{R}^d$ for some $d \geq 1$.

Let p denote the density of P with respect to a counting measure (discrete case) or to the Lebesgue measure (continuous case). Denote this measure by η. The Shannon entropy of the distribution P is defined by

$$H(P) = -\int p \log p \, \mathrm{d}\eta = -\sum_i p_i \log p_i$$

if P is discrete [23]. If P is continuous, then a suitably defined "reference" distribution P_0 is introduced and the entropy is defined by

$$H(P) = -\int \log\left\{\frac{dP}{dP_0}\right\} \mathrm{d}P$$

when P is absolutely continuous with respect to P_0, and $H(P) = -\infty$ otherwise. When both P and P_0 admit densities with respect to the Lebesgue measure then

$$H(P) = -\int p(x) \log\left\{\frac{p(x)}{p_0(x)}\right\} \mathrm{d}x, \tag{1.1}$$

where p_0 is the density function of P_0; see [16]. The Shannon entropy $H(P)$ is interpreted as a measure of the amount of uncertainty described by the probability distribution P (with respect to P_0, usually taken as a uniform distribution).

Maximizing the Shannon entropy over the class Γ_d generates a d-dimensional exponential family [3, 16]. Specifically, if there exists a distribution P_* maximizing (1.1) over the class Γ_d, then its density p_* satisfies

$$p_*(x) = \frac{p_0(x) \exp\{\lambda^T s(x)\}}{\int p_0(x) \exp\{\lambda^T s(x)\} \, \mathrm{d}\eta(x)},$$

where $\lambda = (\lambda_1, \ldots, \lambda_d)$ is a vector of Lagrange multipliers and is determined by the moment constraints $E_P(S) = \varsigma$ which define the class Γ_d; see [12]. Suppose, for example, that X is a continuous random variable and that $d = 2$, $s(X) = (X, X^2)$, and $\varsigma = (\mu_1, \mu_2)$ with $\mu_1 \in \mathbb{R}$ and $\mu_2 > \mu_1^2$ (finite). Suppose also that P_0 corresponds to the Lebesgue measure (that is, a uniform distribution). Then the distribution P_* that maximizes (1.1) is normal with mean μ_1 and variance $\sigma^2 = \mu_2 - \mu_1^2$.

From (1.1), it is clear that maximizing the Shannon entropy with respect to P is equivalent to minimizing the Kullback-Leibler divergence between P and the reference distribution P_0. This motivates the so-called *minimum discrimination information principle*; see [15].

Despite its intuitive appeal, the maximum entropy principle is controversial. However, Topsøe [24] provides an interesting decision theoretical justification for the maximum entropy principle. Grünwald and Dawid [9] extend the work of Topsøe by considering a generalized concept of entropy related to the choice of the loss function. They also discuss generalized divergences and the corresponding generalized "exponential" families of distributions. These families do not seem to have been systematically studied, nor have they been widely used for actual data analysis. The purpose of this paper is to study a connection between a well-known family of f-divergences and a class of distributions that arise as predictive distributions in the analysis of exponential families when standard conjugate priors are assumed on the canonical parameter.

Consider, for example, the Student's t distribution which arises as the predictive distribution for normal data when the precision parameter is assumed to have a conjugate (gamma) prior. When the degrees of freedom tend to infinity, we recover the normal distribution. On the other hand, the Student's t distribution can be derived as the solution to an optimization problem involving a family of generalized divergences indexed by a parameter α which is related to the degrees of freedom. As α tends to infinity, this generalized divergence converges to the Kullback-Leibler divergence, which is minimized by the normal distribution.

In this paper, we discuss this and another important example in some detail, and argue that similar results must also hold for other exponential families. The layout of the paper is as follows. In the next section, we briefly review a family of divergences that has been extensively discussed in the literature. Then, in Sect. 3, we work out the connection described above for the case of the Student's t and Pareto Type II distributions. Finally, in Sect. 4 we discuss how these results could be generalized to other exponential families.

2 A Family of Divergences

Consider the following parametrized family of divergences for $\alpha \geq 1$:

$$D_\alpha(p, p_0) = \begin{cases} \frac{\alpha^2}{(\alpha-1)} \left\{1 - \int \left[\frac{p_0(x)}{p(x)}\right]^{1/\alpha} p(x)\, \mathrm{d}\eta(x)\right\} & (\alpha > 1) \\ \int p_0(x) \log\left\{\frac{p_0(x)}{p(x)}\right\} \mathrm{d}\eta(x) & (\alpha = 1), \end{cases} \tag{2.1}$$

which includes the Kullback-Leibler divergence as the limiting case $\alpha \to \infty$. The above family appears to date back to Rényi [22] (see also Csiszár [5, 7], Havrda and Charvat [11], Rathie and Kannappan [21] and Amari [2]) and is a particular case of the general class of f-divergences defined by

$$D^{(f)}(p, p_0) = \int f\left(\frac{p(x)}{p_0(x)}\right) p_0(x)\, \mathrm{d}\eta(x), \tag{2.2}$$

where $f : [0, \infty) \to (-\infty, \infty]$ is a convex function, continuous at 0 and such that $f(1) = 0$ (see [1, 5, 6, 17]).

For the general entropy $H^{(f)}(P) = -D^{(f)}(p, p_0)$, let

$$\mathcal{F}(p) = H^{(f)}(P) + \lambda^T\{E[s(X)] - \varsigma\} + \lambda_0 \left\{\int p(x)\, \mathrm{d}\eta(x) - 1\right\},$$

be the Lagrange function, where the vector λ and the scalar λ_0 are Lagrange multipliers. The generalized maximum entropy distribution P_* subject to the moment constraints, if it exists, has a density p_* satisfying

$$-\dot{f}\left(\frac{p_*(x)}{p_0(x)}\right) + \lambda^T s(x) + \lambda_0 = 0,$$

where $\dot{f}$ denotes the first derivative of f. Then

$$p_*(x) = p_0(x)\, g(\lambda_0 + \lambda^T s(x))$$

with $g(\cdot) = \dot{f}^{-1}(\cdot)$.

The α-divergence (2.1) is obtained when

$$f(x) = f_\alpha(x) \equiv \begin{cases} \frac{\alpha}{\alpha-1}\{(1-x) - \alpha x(x^{-1/\alpha} - 1)\} & \alpha > 1 \\ (x-1) - \log(x) & \alpha = 1, \end{cases}$$

It is derived from the well-known limit[1]

$$\lim_{\alpha \to \infty} \alpha(x^{1/\alpha} - 1) = \log(x). \tag{2.3}$$

The Kullback-Leibler divergence in (1.1) corresponds to the case $f(x) = f_\infty(x) \equiv \lim_{\alpha \to \infty} f_\alpha(x) = (1-x) + x\log(x)$. Thus,

$$H_\alpha(P) = -D_\alpha(p, p_0) \tag{2.4}$$

generalizes the entropy (1.1).

[1] Here used in the form $\log(x) = -\log(1/x)$.

In the next section, we show how the Student's t distribution can be derived as the distribution on $\mathbb{R}$ which maximizes (2.4) over the class of distributions having first- and second-order moments. Similarly, we also show how the Pareto Type II distribution can be obtained as the distribution on $\mathbb{R}_+$ that maximizes (2.4) over the class of distributions having first-order moments.

3 Main Idea

3.1 *Student's t Distribution*

The Student's t distribution has a density with respect to the Lebesgue measure given by

$$\mathrm{St}(x|\mu,\sigma^2,\gamma)=\frac{\Gamma((\gamma+1)/2)}{\sigma\sqrt{\gamma\pi}\,\Gamma(\gamma/2)}\left[1+\frac{1}{\gamma}\frac{(x-\mu)^2}{\sigma^2}\right]^{-(\gamma+1)/2}\qquad(x\in\mathbb{R}),$$

where $\mu \in \mathbb{R}$ is the location parameter, $\sigma > 0$ is the scale parameter and $\gamma > 0$ denotes the degrees of freedom. Provided that $\gamma > 2$, the first two moments exist and are given by $E(X) = \mu$ and $\mathrm{Var}(X) = \gamma\sigma^2/(\gamma - 2)$.

The t distribution is often used as an alternative to the normal distribution in situations where a robust analysis is required, for instance to account for the presence of possible outliers in the data. It can be regarded as a generalization of the normal distribution in the sense that a t distribution converges to a normal distribution as the degrees of freedom tend to infinity.

Let $d = 2$ and $s(X) = (X, X^2)$. We now show that the Student's t distribution arises as a generalized maximum entropy distribution, in the sense that it maximizes the generalized entropy $H_\alpha(P)$ subject to the constraints $E_P(S) = \varsigma$, where $\varsigma = (\mu_1, \mu_2)$ with $\mu_1 \in \mathbb{R}$ and $\mu_2 > \mu_1^2$ (finite). In order to obtain the usual Student's t distribution, we take the density p_0 to be proportional to the Lebesgue measure, that is, $p_0(x) = r$ for all $x \in \mathbb{R}$ $(r > 0)$.

As pointed out above, in the particular case of the α-divergences given by (2.1) we have $f(x) = f_\alpha(x)$, and hence $g(y) = 1/(1 - y/\alpha)^\alpha$. Now, since

$$\int_{-\infty}^{\infty} p_*(x)\,\mathrm{d}x = 1,$$

we must have

$$\int_{-\infty}^{\infty}\frac{r\,\mathrm{d}x}{\left(1-\frac{\lambda_0+\lambda_1 x+\lambda_2 x^2}{\alpha}\right)^\alpha}=1.$$

Now, setting $\lambda_0 = -(\gamma - 1)/2 - \mu^2/\sigma^2$, $\lambda_1 = 2\mu/\sigma^2$, and $\lambda_2 = -1/\sigma^2$, we get

$$\left(\frac{2\gamma}{\gamma+1}\right)^{(\gamma+1)/2} \int_{-\infty}^{\infty} r \left[1 + \frac{1}{\gamma}\left(\frac{x-\mu}{\sigma}\right)^2\right]^{-(\gamma+1)/2} \mathrm{d}x = 1,$$

where $\gamma = 2\alpha - 1$ (hence, we must have $\alpha > 3/2$ for $E(X)$ and $\mathrm{Var}(X)$ to exist). Finally, we also require that

$$r = \frac{\Gamma((\gamma+1)/2)}{\sigma\sqrt{\gamma\pi}\,\Gamma(\gamma/2)} \left(\frac{\gamma+1}{2\gamma}\right)^{(\gamma+1)/2}.$$

Therefore, we have shown that

$$p_*(x) = \mathrm{St}(x|\mu, \sigma^2, \gamma),$$

that is, a Student's t density with location parameter μ, scale parameter σ and degrees of freedom γ, and such that $\mu = \mu_1$ and $\mu_2 - \mu_1^2 = \gamma\sigma^2/(\gamma - 2)$. An alternative derivation of this result can be found in [8].

Now recall that the density function of a Student's t distribution can be written as

$$\mathrm{St}(x|\mu, \sigma^2, \gamma) = \int_0^{\infty} \mathrm{N}(x|\mu, \sigma^2/y)\,\mathrm{Ga}(y|\gamma/2, \gamma/2)\,\mathrm{d}y.$$

This expression shows that the t distribution corresponds to the predictive distribution for normal data when the precision parameter is assumed to have a conjugate (gamma) prior; see, for example, [4, Section 3.2.2]. As $\gamma \to \infty$, the gamma distribution degenerates at its expected value of 1 and $\mathrm{St}(x|\mu, \sigma^2, \gamma) \to \mathrm{N}(x|\mu, \sigma^2)$. We close this example by noting that $\mathrm{N}(x|\mu, \sigma^2)$ maximizes $H(P)$ subject to the same constraints, namely $E_P(S) = \varsigma$.

3.2 *Pareto Type II Distribution*

The Gamma-Gamma distribution has a density with respect to the Lebesgue measure given by

$$\mathrm{Gg}(x|\gamma, \beta, \nu) = \frac{\beta^{\gamma}}{\Gamma(\gamma)} \frac{\Gamma(\gamma+\nu)}{\Gamma(\nu)} \frac{x^{\nu}}{(\beta+x)^{\gamma+\nu}} \quad (x > 0),$$

where $\gamma > 0$, $\beta > 0$ and $\nu = 0, 1, 2, \ldots$ Provided that $\gamma > 2$, the first two moments exist and are given by

$$E(X) = \nu \frac{\beta}{(\gamma - 1)} \quad \text{and} \quad \mathrm{Var}(X) = \frac{\beta^2[\nu^2 + \nu(\gamma - 1)]}{(\gamma - 1)^2(\gamma - 2)}.$$

This distribution corresponds to the predictive distribution of a gamma sampling model when the scale parameter is assumed to have a conjugate (gamma) prior; see [4, Section 3.2.2].

The Exponential-Gamma distribution (also known as the Pareto Type II distribution) is the particular case where $\nu = 1$. Let $\mu = \beta/\gamma$, then

$$\mathrm{Eg}(x|\gamma, \gamma\mu) \equiv \mathrm{Gg}(x|\gamma, \gamma\mu, 1) = \left[\mu\left(1 + \frac{x}{\mu}\right)\right]^{-(\gamma+1)}.$$

In this case

$$E(X) = \frac{\gamma\mu}{(\gamma - 1)} \quad \text{and} \quad \mathrm{Var}(X) = \frac{\gamma^3\mu^2}{(\gamma - 1)^2(\gamma - 2)}.$$

This distribution can be expressed as a continuous scale mixture of exponential distributions as follows

$$\mathrm{Eg}(x|\gamma, \gamma\mu) = \int_0^\infty \mathrm{Exp}(x|\mu/y)\,\mathrm{Ga}(y|\gamma, \gamma)\,\mathrm{d}y,$$

where $\mathrm{Exp}(x|\mu)$ denotes the density of an exponential distribution with mean μ. As $\gamma \to \infty$, the Gamma distribution degenerates at its expected value of 1 and $\mathrm{Eg}(x|\gamma, \gamma\mu) \to \mathrm{Exp}(x|\mu)$.

The Pareto Type II distribution can thus be regarded as a generalization of the exponential distribution in this sense. It can be used in situations where a more flexible model is required, for instance to allow for overdispersion or to account for outliers in the data.

It is convenient to recall at this point that the exponential distribution is the maximum entropy distribution over the class of continuous distribution on $\mathbb{R}_+$ having first-order moments. Let $d = 1$ and $s(X) = X$. We now show that the Pareto Type II distribution arises as a generalized maximum entropy distribution. Specifically, it maximizes the generalized entropy $H_\alpha(P)$ subject to the constraint $E_P(S) = \mu$, where $\mu \in \mathbb{R}_+$ (finite). As before, we take the density p_0 to be proportional to the Lebesgue measure, so that $p_0(x) = r$ for all $x \in \mathbb{R}_+$ $(r > 0)$.

Since $g(y) = 1/(1 - y/\alpha)^\alpha$ and

$$\int_0^\infty p_*(x)\,\mathrm{d}x = 1,$$

we must have

$$\int_0^\infty \frac{r\,\mathrm{d}x}{\left(1 - \frac{\lambda_0+\lambda_1 x}{\alpha}\right)^\alpha} = 1.$$

Now, setting $\lambda_0 = 0$ and $\lambda_1 = -(\gamma + 1)/\mu$, we get

$$\int_0^\infty \left(1 + \frac{x}{\mu}\right)^{-(\gamma+1)} = 1,$$

where $\gamma = \alpha - 1$. Finally, we also require that $r = 1/\mu$.

We have then shown that

$$p_*(x) = \mathrm{Eg}(x|\gamma, \gamma\mu),$$

which, as pointed out above, tends to $\mathrm{Exp}(x|\mu)$ as $\gamma \to \infty$. We close this example by noting that $\mathrm{Exp}(x|\mu)$ maximizes $H(P)$ subject to the same constraint, namely $E_P(S) = \mu$.

4 Discussion

These results open the door to the development of similar methods of analysis for related or more general cases. For example, it is well-known that the Poisson distribution is not the maximum entropy distribution over the class of distributions on the non-negative integers (the geometric distribution claiming this property). Nonetheless, the Poisson distribution is the maximum entropy distribution over the class of ultra log-concave distributions on the non-negative integers [14], so we can expect the Poisson-Gamma distribution to maximize the generalized entropy (2.4) over the same class. A similar result can be expected to hold in the Binomial case [10].

The predictive distributions considered in this paper are related to the q-exponential families introduced by Naudts [19]; see also [20, Chap. 7]. These families are based on the so-called deformed logarithmic function (inspired by expressions such as (2.3)) and the corresponding (suitably defined) inverse, the deformed exponential function, which can also be used in the definition of the generalized entropy (2.4). Such generalized "exponential" families share some of the nice properties of standard exponential families, but unfortunately are not as tractable. These deformed logarithmic and exponential functions can be further generalized, and one may be able to define the f-divergence (2.2) in terms of them; see [18, 20].

Conjugate predictive distributions can be seen as flexible, robust versions of the corresponding exponential family sampling models. The former seem to share an important optimality property with the latter, and in that sense can be regarded as generalized exponential families. We conjecture that conjugate predictive distributions share other properties and may inherit some of the tractability of standard exponential families, including the existence of useful conjugate priors. Last, but not least, the mixture representation of these predictive distributions should lead to relatively simple analyses via sampling-based methods (such as the Gibbs sampler) when used as sampling models for data analysis.

Acknowledgments We are grateful to an anonymous reviewer for his/her comments, which helped us improve the presentation of this paper. This work was partially supported by *Sistema Nacional de Investigadores*. The second author also wishes to thank Asociación Mexicana de la Cultura A.C.

References

1. Ali, S., Silvey, S.D.: A general class of coefficients of divergence of one distribution from another. J. R. Stat. Soc. B **28**, 131–142 (1966)
2. Amari, S.: Differential-Geometrical Methods in Statistics. Springer, Berlin (1990)
3. Barndorff-Nielsen, O.: Information and Exponential Families in Statistical Theory. Wiley, Chichester (1978)
4. Bernardo, J.M., Smith, A.F.M.: Bayesian Theory. Wiley, Chichester (2000)
5. Csiszár, I.: Eine informationstheoretische Ungleichung und ihre Anwendung auf den Beweis der Ergodizität von Markoffschen Ketten. Publ. Math. Inst. Hungar. Acad. Sci. **8**, 85–107 (1963)
6. Csiszár, I.: Information-type measures of difference of probability distributions and indirect observations. Stud. Sci. Math. Hungar. **2**, 299–318 (1967)
7. Csiszár, I.: Axiomatic characterizations of information measures. Entropy **10**, 261–273 (2008)
8. de Souza, A.M.C., Tsallis, C.: Student's t- and r-distributions: Unified derivation from an entropic variational principle. Physica A **236**, 52–57 (1997)
9. Grünwald, P.D., Dawid, A.P.: Game theory, maximum entropy, minimum discrepancy, and robust Bayesian decision theory. Ann. Stat. **32**, 1367–1433 (2004)
10. Harremoës, P.: Binomial and Poisson distributions as maximum entropy distributions. IEEE Trans. Inf. Theory **47**, 2039–2041 (2001)
11. Havrda, J., Charvat, F.: Quantification method of classification processes, the concept of structural α-entropy. Kybernetica **3**, 30–35 (1967)
12. Jaynes, E.T.: Prior probabilities. IEEE Trans. Syst. Sci. Cybern. **4**, 227–241 (1968)
13. Jaynes, E.T.: Papers on Probability, Statistics and Statistical Physics, 2nd edn. Kluwer Academic Publishers, London (1989)
14. Johnson, O.: Log-concavity and the maximum entropy property of the Poisson distribution. Stoch. Process. Appl. **117**, 791–802 (2007)
15. Kesavan, H.K., Kapur, J.N.: The generalized maximum entropy principle. IEEE Trans. Syst. Man Cybern. **19**, 1042–1052 (1989)
16. Kullback, S.: Information Theory and Statistics. Wiley, New York (1959)
17. Liese, F., Vajda, I.: Convex Statistical Distances. Teuber-texte zur Mathematik, Leipzig (1987)
18. Naudts, J.: Estimators, escort probabilities, and ϕ-exponential families in statistical physics. J. Inequal. Pure Appl. Math. **5**, 102 (2004)
19. Naudts, J.: The q-exponential family in statistical physics. Central Eur. J. Phys. **7**, 405–413 (2009)
20. Naudts, J.: Generalised Thermostatistics. Springer, Berlin (2011)

21. Rathie, P.N., Kannappan, P.: A directed-divergence function of type β. Inform. Contr. **20**, 38–45 (1972)
22. Rényi, A.: On measures of entropy and information. In: Proc. 4th Berkeley Symp. Math. Statist. Prob., pp. 547–561 (1961)
23. Shannon, C.E.: A mathematical theory of communication. Bell Syst. Technical J. **27**, 379–423, (1948)
24. Topsøe, F.: Information theoretical optimization techniques. Kybernetika **15**, 8–27 (1979)

Approximation and Mean Field Control of Systems of Large Populations

Carmen G. Higuera-Chan

Abstract We deal with a class of discrete-time stochastic controlled systems composed by a large population of N interacting individuals. Given that N is large and the cost function is possibly unbounded, the problem is studied by means of a limit model $\mathcal{M}$, known as the mean field model, which is obtained as limit as $N \to \infty$ of the model $\mathcal{M}_N$ corresponding to the system of N individuals in combination with an approximate algorithm for the cost function.

Keywords Systems of interacting individuals · Mean field theory · Approximation algorithm · Discounted criterion

1 Introduction

In this work, we consider a class of discrete-time controlled stochastic systems composed by a population of N interacting individuals. Each individual is classified according to the category where it is located. Let S be the set of categories, which is assumed to be a countable set, and $X_n^N(t)$ be the category of the n-th individual at time t. Moreover, we consider that the evolution of $X_n^N(t)$ is determined by a conditional distribution, defined as follows. For $i, j \in S$ and a_t is the action (or control) selected by a central controller

$$K_{ij}(a) := Pr[X_n^N(t+1) = j | X_n^N(t) = i, a_t = a] \;\; t = 0, 1, \ldots.$$

Additionally, at each stage, a cost is incurred as a consequence of the movement of the individuals in the population. We propose a Markov control model to study this kind of systems and its optimality according to the discounted criterion.

C. G. Higuera-Chan (✉)
Departamento de Matemáticas, Universidad de Sonora, Hermosillo, Sonora, Mexico
e-mail: carhiguera@mat.uson.mx

D. Hernández-Hernández et al. (eds.), *Advances in Probability and Mathematical Statistics*, Progress in Probability 79, https://doi.org/10.1007/978-3-030-85325-9_7

This class of controlled models has been studied, for example, in [2, 8–10, 12]. In fact, this paper can be seen as a sequel to [9]. The essential characteristic in these works is the limitation imposed by the fact N is large. Thinking about this point, the analysis of each individual is complex, and for this reason our approach is directed to the context of the mean field theory. In general terms, let $\mathcal{M}_N$ be the model of the system conformed by N individuals, where the states are conformed by the proportions of individuals in each category. Then, the mean field theory enables to set up an approximating model $\mathcal{M}$ obtained as a limit of $\mathcal{M}_N$ as $N \to \infty$ where the states are the proportions corresponding to the limit of such proportions. The model $\mathcal{M}$, called the mean field control model, results in a deterministic system and independent of N, characterizations that make this model easier to analyze than the model $\mathcal{M}_N$. Further, an optimal policy π_* associated with $\mathcal{M}$ will have a good performance on the original model $\mathcal{M}_N$.

We allow for the cost function r to be unbounded. In works [2, 8–10, 12], authors have been developed their results assuming bounded costs. This fact has allowed to apply fixed point theorems to obtain solutions of the optimality equation, which is not possible in the scenario of our work. This leads to an extra challenge in this problem. To handle this, we introduce an approximation algorithm, defined by a sequence of bounded costs $\{r^l\}$ converging to r. This algorithm is applied in the mean field model $\mathcal{M}$ and is being used to characterize the optimal value v_* and the optimal policies. Finally, we measure its optimal performance in the control model $\mathcal{M}_N$. The key point in our procedure is to adapt the result in [4, 5, 13] for standard Markov control processes to the mean field theory.

The paper is organized as follows. In Sect. 2, we introduce the Markov control model associated with the system of N individuals, as well as the corresponding optimal control problem. Next, in Sect. 3, we define the mean field control model $\mathcal{M}$, the approximation algorithm, and our main results. The proofs are provided in Sect. 4.

Notation

The set $\mathbb{N}$ represents the positive integers, $\mathbb{N}_0 := \mathbb{N} \cup \{0\}$, and $\mathbb{R}$ denotes the set of real numbers.

A Borel space is a Borel subset of a complete and separable metric space. For a Borel space Z, the Borel σ-algebra is denoted by $\mathcal{B}(Z)$, and "measurable", for sets and functions, means "Borel measurable".

The set $\mathbb{B}(Z)$ denotes the class of real-valued bounded functions on Z endowed with the supremum norm $\|h\| := \sup_{z \in Z} |h(z)|$, similarly the set $\mathcal{C}(Z)$ denotes the class of real-valued continuous functions, and the set $\mathcal{C}_b(Z) \subset \mathbb{B}(Z)$ is the subspace of all real-valued bounded and continuous functions on Z.

In addition, $\mathbb{P}(Z)$ is the set of all probability measures on Z, which is also a Borel space. For instance, if Z has the form $Z = \{z_1, z_2, \ldots\}$, a probability measure $p \in \mathbb{P}(Z)$ is identified by $p := \{p(z_i)\}_{i \in \mathbb{N}}$, where $p(z_i) \geq 0$, $\forall i \in \mathbb{N}$ and $\sum_{i=1}^{\infty} p(z_i) =$

1. Furthermore, $\|\cdot\|_\infty$ denotes the corresponding L_∞-norm, that is, for each vector $p \in \mathbb{P}(Z)$:

$$\|p\|_\infty := \sup_{i \in \mathbb{N}} \{|p(z_i)|\}.$$

Let Z and Y be Borel spaces. A stochastic kernel $Q(\cdot|\cdot)$ on Z given Y, is a function $Q : \mathcal{B}(Z) \times Y \to [0,1]$, such that $Q(\cdot|y) \in \mathbb{P}(Z)$ for each $y \in Y$, and $Q(D|\cdot)$ is a measurable function on Y for each $D \in \mathcal{B}(Z)$. We denote by $\mathbb{P}(Z|Y)$ the set of all stochastic kernels on Z given Y.

Finally, 1_D denotes the indicator function of the set D.

2 The N-Objects Markov Control Model

Specifically, we consider a discrete-time controlled system composed by a large population of interacting individuals. The population consists of N ($N \sim \infty$) individuals which can be classified according to a set of categories at each time denoted by $S := \{c_1, c_2, \ldots\} = \{1, 2, \ldots\}$. Let $X_n^N(t)$, $n = 1, \ldots, N$, be the category of the n-th individual at time t. In this sense $X_n^N(t) \in S$. There is a central controller that at each time t picks an action a_t from a Borel set A, producing a random movement among categories for each individual in the population. This movement is determined by a transition probability, homogeneous in N and n, of the form

$$P[X_n^N(t+1) = j | X_n^N(t) = i, a_t = a] = K_{ij}(a), \; i, j \in S, a \in A. \tag{2.1}$$

The relation $K_{ij}(a)$ represents the probability for any object goes from category i to category j when the controller picks the action $a \in A$. Thus, we denote by $K(a) = [K_{ij}(a)]$ the transition matrix corresponding to the action $a \in A$.

Because N is large, the analysis of each individual is practically infeasible. So throughout this work it is assumed that individuals are observable through their current categories, that is, the controller only observes the number of individuals in each of the categories. In this sense, the behavior of the system will be analyzed by means of the proportion of individuals at each category $i \in S$ at time t defined as

$$M_i^N(t) := \frac{1}{N} \sum_{n=1}^{N} 1_{\{X_n^N(t) = i\}},$$

and we denote $M^N(t) = \{M_i^N(t)\}_{i \in S}$ be the empirical distribution of the population over S.

Observe that $M^N(t)$ belongs to the set $\mathbb{P}_N := \{p \in \mathbb{P}(S) : Np(i) \in \mathbb{N}_0, \; \forall i \in S\}$ and represents the state of the system at time t. The dynamic of $M^N(t)$ depends on particular properties of the population, but it is possible to obtain a stochastic difference equation which defines its evolution as in [2, 3]. Specifically, in [3, 11] it is proved that the process $\{M^N(t)\}$ form a no-homogeneous Markov chain. So applying Markov chain simulation techniques it is possible to obtain a function $H^N : \mathbb{P}_N \times A \times \mathbb{R}^N \to \mathbb{P}_N$ such that

$$M^N(t+1) = H^N(M^N(t), a_t, w_t), \ t \in \mathbb{N}_0, \tag{2.2}$$

where $\{w_t\}$ is a sequence of independent and identically distributed (i.i.d.) random vectors on $[0,1]^N$, with common distribution θ defined as follows. For $w = (w_1, w_2, \ldots, w_N) \in [0,1]^N$,

$$\theta(B) = P[w \in B], \ B \in [0,1]^N$$

where each component $w_i \sim Unif[0,1], i = 1, \ldots, N$.

In the remainder, we assume that the dynamic of the process $\{M^N(t)\}$ (see (2.2)) is given by the function H^N as in (2.6). To define this function, we use the transition matrix $K(a)$ and simulate the dynamic of an arbitrary individual of the population. For each $k, j \in S$ and $a \in A$ let

$$\Delta_{kj}(a) := [\Psi_{k,j}(a), \Psi_{k,j+1}(a)] \subseteq [0,1], \tag{2.3}$$

where

$$\Psi_{k,j}(a) := \sum_{d=1}^{j-1} K_{kd}(a), \ k, j \in S. \tag{2.4}$$

Observe that for each $k \in S$ fix, $\{\Delta_{kj}(a)\}_{j \in S}$ is a partition of $[0,1]$ and the length of $\Delta_{kj}(a)$ represents the probability that an individual moves from k to j when the controller takes the action $a \in A$.

Now, for each $t \in \mathbb{N}_0$ we define $\{w_n^k(t)\}$, $n = 1, 2, \ldots, N$, $k \in S$, a family of i.i.d. random variables uniformly distributed on $[0,1]$, which are organized according to the next form

$$w_t := \{w^i(t)\}_{i \in S},$$

$$w^i(t) := (w_1^i(t), \cdots, w^i_{NM_i^N(t)}(t)), \ i \in S.$$

Thus, the proportion of individuals at time $t+1$ in category j takes the form

$$M_j^N(t+1) := \frac{1}{N} \sum_{k=1}^{\infty} \sum_{n=1}^{NM_k^N(t)} 1_{\Delta_{kj}(a_t)}(w_n^k(t)), \ j \in S. \tag{2.5}$$

Therefore, by (2.5) the function H^N takes the form

$$H^N(m,a,w) = \{H_i^N(m,a,w)\}_{i\in S} \tag{2.6}$$

for $(m,a,w) \in \mathbb{P}_N \times A \times [0,1]^N$, where

$$H_j^N(t+1) := \frac{1}{N}\sum_{k=1}^{\infty}\sum_{n=1}^{Nm_k} 1_{\Delta_{kj}(a)}(w_n^k),$$

$m = \{m_i\}_{i\in S}$.

Finally, for a state $M^N(t)$ the controller incurs a cost $r(M^N(t))$. This cost can be represented by a measurable function

$$r : \mathbb{P}(S) \to \mathbb{R}. \tag{2.7}$$

2.1 Formulation of the N-Markov Control Model (N-MCM)

We define the discrete-time Markov control model, associated with the system of N individuals introduced above (N-MCM), as follows:

$$\mathcal{M}_N := \left(\mathbb{P}_N, A, H^N, \theta, r\right).$$

The model $\mathcal{M}_N$ describes the evolution of a system over time. Indeed, at time $t \in \mathbb{N}_0$, the controller observes the state $m = M^N(t) \in \mathbb{P}_N$ which represents the configuration of the population. Next, he/she picks an action $a = a_t \in A$. As a consequence, a cost $r(m)$ incurred, and the system moves to a new state $m' = M^N(t+1)$ according to the transition law

$$\begin{aligned} Q(D|m,a) &:= P\left[M^N(t+1) \in D | M^N(t) = m, a_t = a\right] \\ &= \int_{[0,1]^N} 1_D\left[H^N(m,a,w)\right]\theta(dw),\ D \in \mathcal{B}(\mathbb{P}_N), \end{aligned} \tag{2.8}$$

with H^N as in (2.2). Once the state is m', the procedure is repeated, and the performance of the cost flow $\{r(M^N(t))\}$ is measured using an infinite horizon discounted criterion. Therefore the actions picked by the controller are directed to minimize his/her cost flow.

We impose the following continuity and compactness conditions on the model $\mathcal{M}_N$.

Assumption 2.1

(a) The action space A is a compact Borel space.
(b) The mapping $a \longmapsto K_{ij}(a)$ defined in (2.1) is continuous, for all $i, j \in S$.
(c) For each $j \in S$ and $a \in A$, there exists a finite subset $S_j(a) \subseteq S$ such that $K_{ij}(a) > 0$ for all $i \in S_j(a)$ and $K_{ij}(a) = 0$ for $i \notin S_j(a)$. Moreover,

$$R := \sup_{a \in A} \sup_{j \in S} \#(S_j(a)) < \infty.$$

(d) The one-stage cost r is bounded below and uniformly Lipschitz function with constant L_r; that is, for every $m, m' \in \mathbb{P}(S)$,

$$|r(m) - r(m')| \leq L_r \left\| m - m' \right\|_\infty .$$

Remark 2.2 Observe that Assumptions (a), (b) and (d) are used in [2, 3, 9, 10] and, they are standard in stochastic control theory (see [6, 7]). On the other hand, Assumption 2.1(c) can be interpreted as follows: once the controller takes an action $a \in A$, an individual can move in one step to a category $j \in S$ if it belongs to a category in $S_j(a)$. Another interpretation is that once the controller takes an action a, a generic individual of the population can move to a category $j \in S$ only if the individual belongs to a finite number of categories $S_j(a)$. So, it is not possible to move to a category $j \in S$ from and state in one movement.

Now, we introduce the missing elements to define the optimal control problem for a population with N individuals. Let $\mathbb{H}_0^N := \mathbb{P}_N$ and $\mathbb{H}_t^N := (\mathbb{P}_N \times A)^t \times \mathbb{P}_N$, $t \geq 1$. An element h_t^N of $\mathbb{H}_t^N$ takes the form

$$h_t^N = (M^N(0), a_0, \ldots, M^N(t-1), a_{t-1}, M^N(t)),$$

representing the history of the control system up to time t. A *policy* for the controller is a sequence $\pi^N = \{\pi_t^N\}$ of stochastic kernels $\pi_t^N \in \mathbb{P}(A|\mathbb{H}_t^N)$ such that $\pi_t^N(A|h_t^N) = 1 \ \forall h_t^N \in \mathbb{H}_t^N$, $t \in \mathbb{N}_0$. We denote by Π^N the family of the strategies.

We denote by $\mathbb{F}$ the class of all measurable functions $f : \mathbb{P}(S) \to A$. In addition we define $\mathbb{F}^N := \mathbb{F}|_{\mathbb{P}_N}$ and a policy $\pi^N = \{\pi_t^N\} \in \Pi^N$ is said to be a *Markov policy* if there exists $f_t^N \in \mathbb{F}^N$ such that $\pi_t^N(\cdot|h_t^N) = f_t^N(\cdot|M^N(t))$ for every $t \in \mathbb{N}_0$ and $h_t^N \in \mathbb{H}_t^N$. In this case π^N takes the form $\pi^N = \{f_t^N\}$. The class of all Markov strategies is denoted by Π_M^N. In particular, if $f_t^N \equiv f^N$ for some $f^N \in \mathbb{F}^N$ and for all $t \in \mathbb{N}_0$, we say that π^N is a *stationary* policy.

Remark 2.3 Let Π_M be the sets of Markov policies, when we use $\mathbb{F}$ instead of $\mathbb{F}^N$. Observe that for a policy $\pi = \{f_t\} \in \Pi_M$, if f_t is restricted to $\mathbb{P}_N$, π turns out to be an element of Π^N.

2.2 Optimality in the N-MCM

We fix the discount factor $\alpha \in (0, 1)$. Thus, for each policy $\pi^N \in \Pi^N$ and initial state $m^N(0) = m \in \mathbb{P}_N$, we define the total expected discounted cost as

$$V^N(\pi^N, m) := E_m^{\pi^N} \sum_{t=0}^{\infty} \alpha^t r(M^N(t)),$$

where $E_m^{\pi^N}$ denotes the expectation operator with respect to the probability measure $P_m^{\pi^N}$ induced by the policy π^N and the initial state m. See, for instance, [1] for the construction of $P_m^{\pi^N}$.

Remark 2.4

(a) For each $\pi^N \in \Pi^N$ and $M^N(0) = m \in \mathbb{P}_N$, there exist a probability space $(\Omega', \mathcal{F}', P_m^{\pi^N})$ with $P_m^{\pi^N}$ satisfying

(i) $P_m^{\pi^N}(M^N(0) \in B) = \delta_m(B), \ B \in \mathcal{B}(\mathbb{P}_N)$
(ii) $P_m^{\pi^N}(a_t \in D|h_t^N) = \pi_t^N(D|h_t^N), \ D \in \mathcal{B}(A)$
(iii) Markov-like property

$$\begin{aligned} P_m^{\pi^N}\left[m^N(t+1) \in D|h_t^N, a_t\right] &= Q(D|m^N(t), a_t) \\ &= \int_{[0,1]^N} 1_D\left[H^N\left(m^N(t), a_t, w\right)\right]\theta(dw), \ D \in \mathcal{B}(\mathbb{P}_N), \end{aligned} \tag{2.9}$$

where Q is the transition law in (2.8).

(b) As a consequence of Assumption 2.1, taking into account the structure of function H^N (see (2.5)-(2.6)), it is easy to prove that the mapping $(m, a) \longmapsto H^N(m, a, w)$ is continuous for $w \in [0, 1]^N$. Further, following standard arguments (see, e.g., [7]) and (2.9) the function

$$(m, a) \mapsto \int_{\mathbb{P}_N} u(m')Q(dm'|m, a) = \int_{[0,1]^N} u[H^N(m, a, \theta)]\theta(dw)$$

is continuous for each bounded and continuous function $u : \mathbb{P}_N \to \mathbb{R}$.

Hence, the objective for the controller is to find a policy $\pi_*^N \in \Pi^N$ such that

$$V_*^N(m) := \inf_{\pi^N \in \Pi^N} V^N(\pi^N, m) = V^N(\pi_*^N, m), \ m \in \mathbb{P}_N. \tag{2.10}$$

In this case π_*^N is called the optimal policy, whereas V_*^N is the N-value function.

Remark 2.5 (Sufficiency of Markov Strategies) It is well-known that (see, e.g., [7]) for each $\pi^N \in \Pi^N$ there is a policy $\psi^N \in \Pi_M^N$ such that

$$V^N(\psi^N, \cdot) \leq V^N(\pi^N, \cdot).$$

Then, we can restrict our analysis to the set of Markov policies Π_M^N.

Once defined, the control problem in the N-MCM $\mathcal{M}_N$, we state the following result, Proposition 2.7 below, that provide a characterization of the optimality of the N-values function and the optimal policies (for references, see [6, 14]).

Assumption 2.6 There exists $\pi^N \in \Pi_M^N$ such that $V^N(\pi^N, m) < \infty, m \in \mathbb{P}_N$.

For $u \in C(\mathbb{P}(S))$, $m \in \mathbb{P}(S)$ and $a \in A$ we define

$$\bar{T}u(m) := \min_{a \in A} \left\{ r(m) + \alpha \int_{[0,1]^N} u[H^N(m, a, w)]\theta(w) \right\}$$

Proposition 2.7 *Suppose that Assumptions 2.1 and 2.6 hold. Then:*

(a) V_*^N *is the minimal solution in* $C(\mathbb{P}_N)$ *satisfying the optimality equation*

$$V_*^N(m) = \bar{T} V_*^N(m), \ m \in \mathbb{P}_N \tag{2.11}$$

(b) There exist $f_*^N \in \mathbb{F}^N$, *such that* $f_*^N(m)$ *attains the minimum in (2.11), i.e.,*

$$V_*^N(m) := r(m) + \alpha \int_{[0,1]^N} V_*^N \left[H^N \left(m, f_*^N, w \right) \right] \theta(dw), \ m \in \mathbb{P}_N, \tag{2.12}$$

or, equivalently

$$\Phi^N(m, f_*^N(m)) = 0,$$

where $\Phi^N : \mathbb{P}_N \times A \to \mathbb{R}$ *is the discrepancy function defined as*

$$\Phi^N(m, a) := r(m) + \alpha \int_{[0,1]^N} V_*^N[H^N(m, a, w)]\theta(dw) - V_*^N(m). \tag{2.13}$$

Moreover, the policy $\pi_*^N = \{f_*^N\} \in \Pi_M^N$ *is optimal.*

Observe that in the case that r is continuous and bounded function, applying the Banach fixed point Theorem (see [6, 7, 14]) it can be proved that the corresponding optimal value function is the unique solution bounded and continuous of the optimality equation, i.e., $V_*^N \in C_b(\mathbb{P}_N)$. This fact will be used below, in Proposition 3.2. Furthermore, another limitation of Proposition 2.7 is that the optimality equation (2.12) depends on an N-dimensional multiple integral, which, since $N \to \infty$, is practically impossible to handle. These obstacles will be overcome in the next section.

3 The Mean Field Control Model

We consider a controlled deterministic system $\{m(t)\} \in \mathbb{P}(S)$ whose evolution is according to a difference equation

$$m(t+1) = H(m(t), a_t), \tag{3.1}$$

with a given initial condition $m(0) = m \in \mathbb{P}(S)$, where $a_t \in A$ represents the action at time t, and $H : \mathbb{P}(S) \times A \to \mathbb{P}(S)$ is the function defined by

$$H(m, a) := mK(a),$$

where $K(a)$ is the transition matrix of each individual (see (2.1)). Observe that the function H is continuous by the continuity of $K(a)$ in Assumption 2.1(b).

Moreover, because $\{m(t)\} \in \mathbb{P}(S)$ is a deterministic process, we have that the actions are selected by the controller using Markov strategies from the set Π_M, introduced in Remark 2.3.

The convergence of the process $\{M^N(t)\} \in \mathbb{P}_N$ to $\{m(t)\} \in \mathbb{P}(S)$ when the controller takes the same control is deduced from the following result, Theorem 3.1.

Let $\mathcal{E} \neq \emptyset$ be the set of arrays $\varepsilon = \{\varepsilon_{ij}\}_{i,j\in S}$ where ε_{ij} are positive numbers such that

$$\|\varepsilon\|_{\mathcal{E}} := \sup_{j\in S} \sum_{i=1}^{\infty} \varepsilon_{ij} < \infty. \tag{3.2}$$

In effect, $\mathcal{E}$ is not an empty set. An element of $\mathcal{E}$ is, for example, $\varepsilon = \{\frac{1}{(j+1)^i}\}$, because for each $j \in S$ fixed $\sum_{i=1}^{\infty} \frac{1}{(j+1)^i}$ is a geometric series.

Theorem 3.1 *Suppose that $M^N(0) = m(0) = m \in \mathbb{P}_N$. Then, there exist positive constants C and λ, and a function $\mu_T : \mathcal{E} \to \mathbb{R}$, where $\mu_T(\varepsilon) \to 0$ as $\|\varepsilon\|_{\mathcal{E}} \to 0$, such that*

(a) for any $T \in \mathbb{N}$ and $\varepsilon = \{\varepsilon_{ij}\} \in \mathcal{E}$,

$$\sup_{\pi\in\Pi_M} P_m^{\pi} \left\{ \sup_{0\leq t\leq T} \left\| M^N(t) - m(t) \right\|_{\infty} > \mu_T(\varepsilon) \right\} \leq CTe^{-\lambda N\varepsilon_{ij}^2} \; \forall i, j \in S.$$

(b) Furthermore, for each $m \in \mathbb{P}_N$, $T \in \mathbb{N}$, and $\varepsilon = \{\varepsilon_{ij}\} \in \mathcal{E}$,

$$\sup_{\pi\in\Pi_M} E_m^{\pi} \left[\sup_{0\leq t\leq T} \|M^N(t) - m(t)\|_{\infty} \right] \leq CTe^{-\lambda N\varepsilon_{ij}^2} + \mu_T(\varepsilon) \; \forall i, j \in S. \tag{3.3}$$

For its proof, see Sect. 4 Proofs.

Accordingly to previous description, we introduce the following deterministic control model corresponding to the state process (3.1):

$$\mathcal{M} = (\mathbb{P}(S), A, H, r),$$

where r is the one-stage cost function in (2.7), which we call the mean field control model.

Now, for each strategy $\pi \in \Pi_M$ and initial condition $m(0) = m \in \mathbb{P}(S)$, we define the total discounted cost in the mean field as

$$v(\pi, m) = \sum_{t=0}^{\infty} \alpha^t r\left(m(t)\right),$$

and the associated mean field optimal control problem is to find a policy $\pi_* \in \Pi_M$ such that

$$v_*(m) := \inf_{\pi \in \Pi_M} v(\pi, m) = v_*(\pi_*, m), \;\; m \in \mathbb{P}(S),$$

where v_* is the mean field value function and π_* is said to be an optimal policy for the mean field control model $\mathcal{M}$.

3.1 Approximation via Bounded Costs

Recall that r is assumed to be a possibly unbounded function, which in turns implies that so is v_*. Clearly, this fact makes the analysis difficult. Hence, we introduce an approximation scheme defined by a sequence of bounded cost functions $\{r^l\}$ increasingly converging to r, that is

$$r^l \nearrow r \text{ as } l \to \infty. \tag{3.4}$$

To fix ideas we set

$$r^l(m) := \min\{r(m), l\}. \tag{3.5}$$

Each r^l is a bounded function and inherits the Lipschitz property from r. Now, for each $l \in \mathbb{N}$ we define the total discount cost in the mean field model associated with the cost function r^l as

$$u_l(\pi, m) := \sum_{t=0}^{\infty} \alpha^t r^l(m(t))$$

and the corresponding mean field mean field control problem is to find a policy $\pi_*^l \in \Pi_M$ such that

$$u_l^*(m) := \inf_{\pi \in \Pi_M} u_l(m, \pi).$$

Under Assumption 2.1 and using the fact that H is a continuous function, it is easy to prove that the dynamic programming operator

$$\mathcal{T}_l h(m) := \inf_{a \in A} \left\{ r^l(m) + \alpha h[H(m,a)] \right\}, \quad m \in \mathbb{P}(S),$$

maps $\mathcal{C}_b(\mathbb{P}(S))$ into itself, that is, $\mathcal{T}_l h \in \mathcal{C}_b(\mathbb{P}(S))$ for $h \in \mathcal{C}_b(\mathbb{P}(S))$. Moreover we have the following result (see [6, 14]).

Proposition 3.2 *Suppose that Assumption 2.1 (a), (b), (c) holds and $\{r^l\}$ as in (3.5). Then:*

(a) For each $l \in \mathbb{N}$, the function u_*^l is the unique solution in $\mathcal{C}_b(\mathbb{P}(S))$ satisfying

$$u_*^l(m) = \mathcal{T}_l u_*^l(m), \quad m \in \mathbb{P}_N.$$

(b) There exist $f_*^l \in \mathbb{F}$ such that for each $m \in \mathbb{P}(S)$,

$$u_*^l(m) := r^l(m, f_*^l) + \alpha v_*^l \left[H\left(m, f_*^l\right) \right]. \tag{3.6}$$

Next, we present the approximation result for v_* in the mean field control model $\mathcal{M}$, which is one of the main problems we addressed.

Assumption 3.3 There exist $\pi \in \Pi_M$ such that $v(\pi, m) < \infty$, $m \in \mathbb{P}(S)$.

Theorem 3.4 *Suppose that Assumptions 2.1 and 3.3 hold.*

(a) The mean field value function v_* is the minimal function in $C(\mathbb{P}(S))$ that satisfies the optimality equation for the control model, i.e.,

$$v_*(m) = \inf_{a \in A} \{ r(m) + \alpha v_*[H(m,a)] \}, \quad m(0) = m \in \mathbb{P}(S). \tag{3.7}$$

(b) There exist $f \in \mathbb{F}$, such that attains the minimum in (3.7). Moreover, the policy $\hat{\pi} = \{f\}$ is optimal for the mean field control problem in $\mathcal{M}$.
(c) As $l \to \infty$, $u_*^l(m) \nearrow v_*(m)$ for $m(0) = m \in \mathbb{P}(S)$;
(d) The policy $\hat{\pi}_* = \{f_*^1, f_*^2, \ldots\}$ where f_*^l satisfy (3.6) for each $l \in \mathbb{N}$ is eventually optimal for the mean field control problem in $\mathcal{M}$. That is, for any initial condition $m(0) = m \in \mathbb{P}(S)$

$$\lim_{l \to \infty} \Phi(m(l), a_l) = 0,$$

where $a_l = f_*^l(m(l))$ *and*

$$\Phi(m, a) := r(m) + \alpha v_*[H(m, a)] - v_*(m).$$

Proof See Sect. 4. □

In what follows, we consider the set of policies

$$\bar{\Pi}_M := \{\pi \in \Pi_M | v(\pi, m) < \infty \text{ for each } m \in \mathbb{P}(S)\},$$

and

$$\bar{\Pi}_M^N := \{\pi \in \Pi_M^N | V^N(\pi, m) < \infty \text{ for each } m \in \mathbb{P}_N\},$$

which are non empty sets by Assumptions 2.6 and 3.3. These assumptions guarantee that the discounted indices are well defined even though the cost function r may not be bounded.

Observe that Theorem 3.4 provides a way to approximate the mean field value function v_* as well as the corresponding optimal policy via the sequences $\{u_*^l\}$ and $\{f_*^l\}$, easily computable from Proposition 3.2. Thus, our objective is to study the behavior of the policy $\hat{\pi}_* = \{f_*^l\}$ in the original control model $\mathcal{M}_N$. In other words, we analyze the optimality deviation in $\mathcal{M}_N$ when $\hat{\pi}_*$ is used to control the original process $\{M^N(t)\}$. Specifically we prove that such optimality deviation vanishes as $N \to \infty$, as long as V_*^N approaches v_*. We state our main result as follows.

Theorem 3.5 *Suppose that Assumptions 2.1, 2.6 and 3.3 hold, and* $M^N(0) = m(0) = m \in \mathbb{P}_N$. *Then:*

(a) For $\varepsilon \in \mathcal{E}$ *(see (3.2)),* $T \in \mathbb{N}$, $0 \le t \le T$, *and* $\pi' \in \Pi_M$,

$$E_m^{\pi'}\left[\left|V_*^N(M^N(t)) - v_*(m(t))\right|\right]$$
$$\le E_m^{\pi'}\bar{\gamma}_T(m(t)) + E_m^{\pi'}\bar{\gamma}_T'(M^N(t)) + L_r\frac{1-\alpha^T}{1-\alpha}\left[CTe^{-\lambda N\varepsilon_{ij}^2} + \mu_T(\varepsilon)\right] \forall i, j \in S,$$

where the constant L_r *comes from Assumption 2.1, while* C, λ *and* $\mu_T(\varepsilon)$ *are as in Theorem 3.1,* $E_m^{\pi'}\bar{\gamma}_T(m(t)) \to 0$ *and* $E_m^{\pi'}\bar{\gamma}_T'(M^N(t)) \to 0$ *as* $T \to \infty$. *Hence,*

$$\lim_{N\to\infty}\left\|V_*^N - v_*\right\| = 0.$$

(b) The policy $\hat{\pi}_* = \{f_*^l\}$ *is eventually asymptotically optimal for control model* $\mathcal{M}_N$, *that is*

$$\lim_{l\to\infty}\lim_{N\to\infty}\Phi^N(m(l), f_*^l(m)) = 0.$$

Proof The proof is given in Sect. 4. □

Remark 3.6 Observe that for each $t \in \mathbb{N}_0$, $m \in \mathbb{P}_N$ and $\pi \in \Pi_M$,

$$\lim_{l\to\infty}\lim_{N\to\infty} E_m^\pi |V_*^N(M^N(t)) - u_*^l(m(t))| = 0. \tag{3.8}$$

Indeed, for each $t \in \mathbb{N}_0$,

$$|V_*^N(M^N(t)) - u_*^l(t)| \leq |V_*^N(M^N(t)) - v_*(m(t))| + |v_*(m(t)) - u_*^l(m(t))|,$$

Hence, (3.8) follows from Theorem 3.4(b) and Theorem 3.5(a).

4 Proofs

4.1 *Proof of Theorem 3.1*

For an arbitrary policy $\pi = \{f_t\} \in \Pi_M$, let $\{a_t\} \in A$ be sequence of actions corresponding to its application, and $m \in \mathbb{P}_N$ be initial condition. We define

$$B_{inj}^N(t) := 1_{\Delta_{ij}(a_t)}(w_n^i(t)), \quad i, j, n \in S, \tag{4.1}$$

where $w_n^i(t)$ are i.i.d.random variables uniformly distributed on [0, 1] (see (2.3)). Hence, for each $t \in \mathbb{N}_0$, $\{B_{inj}^N(t)\}_{inj}$ is a family of i.i.d. Bernoulli random variables with mean

$$E_m^\pi\left[B_{inj}^N(t)|a_t = a\right] = K_{ij}(a).$$

For a fixed $\varepsilon \in \mathcal{E}$ (see 3.2), from Hoeffding's inequality, we have for each $t \in \mathbb{N}_0, i, j \in S$,

$$P_m^\pi\left[\left|\sum_{n=1}^{NM_i^N(t)} B_{inj}^N(t) - NM_i^N(t)K_{ij}(a_t)\right| < N\varepsilon_{ij}\right] > 1 - 2e^{-2N\varepsilon_{ij}^2}. \tag{4.2}$$

Now, let

$$\Omega_{ij} = \left\{w \in \Omega' : \left|\sum_{n=1}^{NM_i^N(t)} B_{inj}^N(t) - NM_i^N(t)K_{ij}(a_t)\right| < N\varepsilon_{ij}\right\} \subset \Omega',$$

and define $\bar{\Omega} = \bigcap_{i,j\in S} \Omega_{ij}$.

On the other hand, we define the sequence $\{\mu_t(\varepsilon)\}_{t\in\mathbb{N}_0}$ as

$$\mu_0(\varepsilon) = 0; \quad \mu_t(\varepsilon) = \|\varepsilon\|_{\mathcal{E}} \sum_{d=0}^{t-1} R^d,$$

where $R := \sup_{a\in A} \sup_{j\in S} \#(S_j(a))$ (see Assumption 2.1(c)). Now we prove that on $\bar{\Omega}$ the following holds

$$\|M^N(t) - m(t)\|_\infty \le \mu_t(\varepsilon) \ \forall t \in \mathbb{N}_0. \tag{4.3}$$

We proceed by induction. For $t = 0$, by assumption, we have $\|M^N(0) - m(0)\|_\infty = 0$. We assume that $\|M^N(k) - m(k)\|_\infty \le \mu_k(\varepsilon)$ for some $k \in \mathbb{N}$. Then from (2.5), (3.1), and (4.1), for each $j \in S$

$$\begin{aligned}
|M_j^N(k+1) - m_j(k+1)| &= \left| \sum_{i=1}^{\infty} \frac{1}{N} \left[\sum_{n=1}^{NM_i^N(k)} B_{inj}^N(k) - Nm_i(k)K_{ij}(a_k) \right] \right| \\
&\le \sum_{i=1}^{\infty} \frac{1}{N} \left| \sum_{n=1}^{NM_i^N(k)} B_{inj}^N(k) - Nm_i(k)K_{ij}(a_k) \right| \\
&\le \sum_{i=1}^{\infty} \frac{1}{N} \left| \sum_{n=1}^{NM_i^N(k)} B_{inj}^N(k) - NM_i^N(k)K_{ij}(a_k) \right| \\
&\quad + \sum_{i=1}^{\infty} |M_i^N(k) - m_i(k)| K_{ij}(a_k) \\
&\le \sum_{i=1}^{\infty} \varepsilon_{ij} + \#(S_j(a_k))\mu_k(\varepsilon) \le \sum_{i=1}^{\infty} \varepsilon_{ij} + R\mu_k(\varepsilon).
\end{aligned}$$

Hence, $|M_j^N(k+1) - m_j(k+1)| \le \sum_{i=1}^{\infty} \varepsilon_{ij} + R\mu_k(\varepsilon) \ \forall k \in \mathbb{N}_0$, on the set $\bar{\Omega}$, and

$$\begin{aligned}
\|M^N(k+1) - m(k+1)\|_\infty &\le \sup_{j\in S} \sum_{i=1}^{\infty} \varepsilon_{ij} + R\mu_k(\varepsilon) \\
&= \|\varepsilon\|_{\mathcal{E}} + R\mu_k(\varepsilon) = \|\varepsilon\|_{\mathcal{E}} + R\|\varepsilon\|_{\mathcal{E}} \sum_{d=0}^{k-1} R^d \\
&= \|\varepsilon\|_{\mathcal{E}} \sum_{d=0}^{k} R^d = \mu_{k+1}(\varepsilon),
\end{aligned}$$

which proves (4.3).

Now, observe that $\{\mu_t(\varepsilon)\}$ form an increasing sequence on t. Thus, for any $T \in \mathbb{N}$, $\mu_t(\varepsilon) \leq \mu_T(\varepsilon)$ $\forall t \leq T$. Therefore, from (4.2), (4.3), and a direct induction over T, under the policy $\pi \in \Pi_M$ we have

$$P_m^\pi \left[\sup_{0 \leq t \leq T} \|M^N(t+1) - m(t+1)\| > \mu_T(\varepsilon) \right] \leq 2Te^{-2N\varepsilon_{ij}^2}\ \forall i, j \in S.$$

Taking $C = \lambda = 2$ we prove part (a).

(b) We denote by $Y_T := \sup_{0 \leq t \leq T} \|M^N(t) - m(t)\|_\infty$. Let $\pi \in \Pi_M$ be an arbitrary policy. For each $m \in \mathbb{P}_N$, $T \in \mathbb{N}$, and $\varepsilon \in \mathcal{E}$, since $Y_T \leq 1$ we have

$$\begin{aligned} E_m^\pi[Y_T] &= E_m^\pi[Y_T 1_{\{Y_T > \mu_T(\varepsilon)\}} + Y_T 1_{\{Y_T \leq \mu_T(\varepsilon)\}}] \\ &\leq P_m^\pi\,(Y_T > \mu_T(\varepsilon)) + \mu_T(\varepsilon) P_m^\pi\,(Y_T \leq \mu_T(\varepsilon)). \end{aligned}$$

Hence from part (a) we obtain

$$\sup_{\pi \in \Pi_M} E_m^\pi[Y_T] \leq CTe^{-\lambda N\varepsilon_{ij}^2} + \mu_T(\varepsilon),\ \forall i, j \in S,$$

that is, part (b) holds. □

4.2 *Proof of Theorem 3.4*

Part (a) and (b) is a well-known result in discounted Markov control processes (see [7]), while part (c) follows from [7, Theorem 4.4.1] (see also [4, 5, 13]).

We proceed to prove part (d). For each $l \in \mathbb{N}_0$ we define the function

$$\Phi_l(m, a) := r^l(m) + \alpha u_*^l[H(m, a)] - u_*^l(m),\ m \in \mathbb{P}(S). \tag{4.4}$$

Observe that $\Phi_l(m, f_*^l(m)) = 0$ (see (3.6)).

On the other hand, for each $m \in \mathbb{P}(S)$ and $n \in \mathbb{N}_0$,

$$\begin{aligned} &\sup_{a \in A} |\Phi(m, a) - \Phi_l(m, a)| \\ &= \sup_{a \in A} |r(m) + \alpha v_*[H(m, a)] - v_*(m) - r^l(m) - \alpha u_l^*[H(m, a)] + u_l^*(m)| \\ &\leq \sup_{a \in A} |\, r(m) - r^l(m)| + \alpha |v_*[H(m, a)] - u_*^l[H(m, a)]| + |v_*(m) - u_l^*(m)| \end{aligned} \tag{4.5}$$

Now, because A is a compact set and the continuity of the functions r, v_*, u^l_*, and H, letting $l \to \infty$ in (4.5), from (3.4) and part (b) we obtain

$$\sup_{a \in A} |\Phi(m, a) - \Phi_l(m, a)| \to 0 \text{ as } l \to \infty, m \in \mathbb{P}(S). \tag{4.6}$$

Finally, we denote $a_l = f^l_*(m(l))$ and observe that

$$|\Phi(m(l), a_l)| = |\Phi(m(l), a_l) - \Phi_l(m(l), a_l)| \leq \sup_{a \in A} |\Phi(m, a) - \Phi_l(m, a)|. \tag{4.7}$$

Taking limit as $l \to \infty$ in (4.7), from (4.6) we get

$$\Phi(m(l), a_l) \to 0 \text{ as } l \to \infty.$$

Therefore $\hat{\pi}_*$ is eventually optimal in the model $\mathcal{M}$.□

4.3 Proof of Theorem 3.5

(a) For an arbitrary policy $\pi' \in \Pi_M$. We denote by $M^N(t) := M^N_{\pi'}(t)$ and $m(t) := m_{\pi'}(t)$ the corresponding trajectories to application of the policy π', with initial condition $M^N(0) = m(0) = m \in \mathbb{P}_N$.

On the other hand, we define the selector

$$\bar{f}(\cdot|m) = f^N_*(\cdot|m) 1_{\mathbb{P}_N}(m) + \tilde{f}(\cdot|m) 1_{[\mathbb{P}_N]^c}(m) \in \mathbb{F}.$$

where f^N_* corresponds to the optimal stationary policy $\pi^N_* = \{f^N_*\}$ in the model $\mathcal{M}_N$ (see Proposition 2.7) and $\tilde{f}$ is an arbitrary selector in $\mathbb{F}$ such that $\tilde{\pi} = \{\tilde{f}\} \in \bar{\Pi}_M$.

From (2.2) and the assumption that $M^N(0) = m(0) = m \in \mathbb{P}_N$, we have that the process $\{M^N(t)\}$ evolves in the set $\mathbb{P}_N$, that is $M^N(t) \in \mathbb{P}_N$ $\forall t \in \mathbb{N}_0$. Thus, from Proposition 2.7, relation (2.10), the definition of $\bar{\pi} \in \Pi_M$, and Remark 2.5, we have

$$\begin{aligned}\inf_{\pi \in \Pi^N_M} V^N(\pi, M^N(t)) &= \inf_{\pi \in \bar{\Pi}^N_M} V^N(\pi, M^N(t)) = V^N_*(M^N(t)) = V^N(\pi^N_*, M^N(t)) \\ &= V^N(\bar{\pi}, M^N(t)) = \inf_{\pi \in \bar{\Pi}_M} V^N(\pi, M^N(t)) \quad P^{\pi'}_m - a.s,\ t \in \mathbb{N}_0,\end{aligned}$$

and for $\hat{\pi} \in \bar{\Pi}_M$ as in Theorem 3.4(b), we have

$$v_*(m(t)) = \inf_{\pi \in \bar{\Pi}_M} v(\pi, m(t)) = v(\hat{\pi}, m(t)).$$

These facts imply

$$V_*^N(M^N(t)) - v_*(m(t)) \geq V^N(\bar{\pi}, M^N(t)) - v(\bar{\pi}, m(t))$$
$$V_*^N(M^N(t)) - v_*(m(t)) \leq V^N(\hat{\pi}, M^N(t)) - v(\hat{\pi}, m(t))$$

and consequently

$$|V_*^N(M^N(t)) - v_*(m(t))|$$
$$\leq \max\{|V^N(\bar{\pi}, M^N(t)) - v(\bar{\pi}, m(t))|, |V^N(\hat{\pi}, M^N(t)) - v(\hat{\pi}, m(t))|\}$$
$$P_m^{\pi'} - a.s, \quad t \in \mathbb{N}_0.$$

With out loss of generality, we suppose that

$$|V_*^N(M^N(t)) - v_*(m(t))| \leq |V^N(\bar{\pi}, M^N(t)) - v(\bar{\pi}, m(t))|. \tag{4.8}$$

Furthermore, for each $m \in \mathbb{P}_N, 0 \leq t < T$, we define V_T^N and v_T as the total discounted cost with finite horizon T, that is, for each $m \in \mathbb{P}_N$

$$V_T^N(\pi, m) := E_m^{\pi}\left[\sum_{k=0}^{T-1} \alpha^k r(M^N(k))\right], \quad v_T(\pi, m) := \sum_{k=0}^{T-1} \alpha^k r(m(k)).$$

Continuing with (4.8)

$$|V_*^N(M^N(t)) - v_*(m(t))| \leq |V^N(\bar{\pi}, M^N(t)) - v(\bar{\pi}, m(t))|$$
$$\leq \left|V^N(\bar{\pi}, M^N(t)) - V_T^N(\bar{\pi}, M^N(t))\right| \tag{4.9}$$
$$+ \left|V_T^N(\bar{\pi}, M^N(t)) - v_T(\bar{\pi}, m(t))\right| \tag{4.10}$$
$$+ |v_T(\bar{\pi}, m(t)) - v(\bar{\pi}, m(t))| \quad P_m^{\pi'} - a.s. \tag{4.11}$$

We now proceed to analyze the terms of in relations (4.9)–(4.11). From Assumption 2.1(c) and (3.3), for an arbitrary $\varepsilon = \{\varepsilon_{ij}\} \in \mathcal{E}$ (see (3.2))

$$E_m^{\bar{\pi}}\left|r(M^N(t)) - r(m(t))\right| \leq L_r E_m^{\bar{\pi}}\left[\|M^N(t) - m(t)\|_\infty\right]$$
$$\leq L_r E_m^{\bar{\pi}}\left[\sup_{0\leq t\leq T} \|M^N(t) - m(t)\|_\infty\right]$$
$$\leq L_r \left(CTe^{-\lambda N\varepsilon_{ij}^2} + \mu_T(\varepsilon)\right) \forall i, j \in S.$$

Hence, for $0 \leq t \leq T$,

$$
\begin{aligned}
&|v(\bar{\pi}, m(t)) - v_T(\bar{\pi}, m(t))| \\
&\qquad \leq \left| \sum_{k=0}^{\infty} \alpha^k r(m(k)) - \sum_{k=0}^{T-1} \alpha^k r(m(k)) \right| \\
&\qquad = \left| \sum_{k=T}^{\infty} \alpha^k r(m(k)) \right| =: \bar{\gamma}_T(m(t)) \;\; P_m^{\pi'} - a.s. \qquad (4.12)
\end{aligned}
$$

where $\bar{\gamma}_T(m(t))$ is a finite function due to $v(\bar{\pi}, m) < \infty$ and is decreasing when $T \to \infty$. Similarly, we can prove for (4.9) that, for $0 \leq t \leq T$, the following bound holds:

$$
\begin{aligned}
&|V_T^N(\bar{\pi}, m(t)) - V^N(\bar{\pi}, m(t))| \\
&\qquad \leq \left| E_{M^N(t)}^{\bar{\pi}} \sum_{k=T}^{\infty} \alpha^k r(M^N(k)) \right| =: \bar{\gamma}_T'(M^N(t)) \;\; P_m^{\pi'} - a.s. \qquad (4.13)
\end{aligned}
$$

We also have, for the term (4.10) the following relations. For $0 \leq t \leq T$,

$$
\begin{aligned}
&|V_T^N(\bar{\pi}, M^N(t)) - v_T(\bar{\pi}, m(t))| \qquad (4.14) \\
&\qquad = \left| E_{M^N(t)}^{\bar{\pi}} \left[\sum_{k=0}^{T-1} \alpha^k r(M^N(k)) - \sum_{k=0}^{T-1} \alpha^k r(m(k)) \right] \right| \\
&\qquad \leq \sum_{k=0}^{T-1} \alpha^k E_{M^N(t)}^{\bar{\pi}} \left| r(M^N(k)) - r(m(k)) \right| \\
&\qquad \leq L_r \frac{1-\alpha^T}{1-\alpha} \left[CTe^{-\lambda N \varepsilon_{ij}^2} + \mu_T(\varepsilon) \right] \; \forall i, j \in S, \;\; P_m^{\pi'} - a.s. \qquad (4.15)
\end{aligned}
$$

Finally, by (4.9)–(4.11), (4.12)–(4.15), and taking expectation $E_m^{\pi'}$, we prove the part (a) of the Theorem 3.5. □

(b) First observe that for all policies $\pi = \{f_t\} \in \Pi_M$, from the Markov property (2.9), and using that $M^N(0) = m(0) = m \in \mathbb{P}_N$, we have

$$
E_m^{\pi} \left[V_*^N \left(M^N(1) \right) \right] = \int_{[0,1]^N} V_*^N \left[H^N(m, f_0, w) \right] \theta(dw).
$$

Now, let $m \in \mathbb{P}(S)$ fixed. Then, from the definition of Φ_l and Φ^N (see (4.4) and (2.13)) we have

$$\begin{aligned}
\Phi^N(m, f_*^l(m)) &= |\Phi^N(m, f_*^l(m)) - \Phi_l(m, f_*^l(m))| \\
&\leq |r(m) - \alpha \int_{[0,1]^N} V_*^N[H^N(m, f_*^l(m), w)]\theta(dw) - V_*^N(m) \\
&- r^l(m) - \alpha u_*^l[H(m,a)] + u_*^l(m)| \\
&\leq |r(m) - r^l(m)| + |V_*^N(m) - u_*^l(m)| \\
&+ |\int_{[0,1]^N} V_*^N[H^N(m, f_*^l(m), w)]\theta(dw) - u_*^l[H(m,a)]| \\
&\leq |r(m) - r^l(m)| + |V_*^N(m) - u_*^l(m)| \\
&+ E_m^{\hat{\pi}_*}|V_*^N(M^N(1)) - u_*^l(m(1))|.
\end{aligned}$$

Hence, letting $N \to \infty$ and $l \to \infty$, (3.4) and Remark 3.6 proves the desired result. □

Acknowledgments Work partially supported by Consejo Nacional de Ciencia y Tecnología (CONACYT-México) under grant Ciencia Frontera 2019-87787.

References

1. Dynkin, E.B., Yushkevich, A.A. Controlled Markov Processes. Springer, New York (1979)
2. Gast, N., Gaujal, B.: A mean field approach for optimization in discrete time. Discrete Event Dyn. Syst. **21**, 63–101 (2011)
3. Gast, N., Gaujal, B., Le Boudec, J.Y.: Mean field for Markov decision processes: from discrete to continuous optimization. IEEE Trans. Autom. Control **57**, 2266–2280 (2012)
4. González-Hernández, J., López-Martínez, R.R., Minjárez-Sosa, J.A.: Approximation, estimation and control of stochastic systems under a randomized discounted cost criterion. Kybernetika **45**(5), 737–754 (2009)
5. Gordienko, E.I., Minjárez-Sosa, J.A.: Adaptive control for discrete-time Markov processes with unbounded costs: discounted criterion. Kybernetika **34**, 217–234 (1998)
6. Hernández-Lerma, O.: Adaptive Markov Control Processes. Springer, New York (1989)
7. Hernández-Lerma, O., Lasserre, J.B.: Discrete-Time Markov Control Processes: Basic Optimality Criteria. Springer, New York (1996)
8. Higuera-Chan, C.G., Minjárez-Sosa, J.A.: A mean field approach for discounted zero-sum games in a class of systems of interacting objects. Dyn. Games Appl. (2021). https://doi.org/10.1007/s13235-021-00377-0
9. Higuera-Chan, C.G., Jasso-Fuentes, H., Minjárez-Sosa, J.A.: Discrete-time control for systems of interacting objects with unknown random disturbance distributions: a mean field approach. Appl. Math. Optim. **74**, 197–227 (2016)
10. Higuera-Chan, C.G., Jasso-Fuentes, H., Minjárez-Sosa, J.A.: Control systems of iteracting objects modeled as a game against nature under a mean field approach. J. Dyn. Games **4**, 59–74 (2017)

11. Le Boudec, J.Y., McDonald, D., Mundinger, J.: A generic mean field convergence result for systems of interacting objects. In: 4th Int. Conf. Quantitative Evaluation of Systems (2007)
12. Martínez-Manzanares, M.E., Minjárez-Sosa, J.A.: A mean field absorbing control model for interacting objects systems. Discrete Event Dyn Syst. (2021). https://doi.org/10.1007/s10626-021-00339-z
13. Minjárez-Sosa, J.A.: Approximation and estimation in Markov control processes under a discounted criterion. Kybernetika **40**(6), 681–690 (2004)
14. Puterman, M.L.: Markov Decision Processes. Discrete Stochastic Dynamic Programming. John Wiley & Sons, New York (1994)

Explicit Solution Simulation Method for the 3/2 Model

Iro René Kouarfate, Michael A. Kouritzin, and Anne MacKay

Abstract An explicit weak solution for the 3/2 stochastic volatility model is obtained and used to develop a simulation algorithm for option pricing purposes. The 3/2 model is a non-affine stochastic volatility model whose variance process is the inverse of a CIR process. This property is exploited here to obtain an explicit weak solution, similarly to Kouritzin (Int J Theor Appl Finance 21(01):1850006, 2018) for the Heston model. A simulation algorithm based on this solution is proposed and tested via numerical examples. The performance of the resulting pricing algorithm is comparable to that of other popular simulation algorithms.

Keywords 3/2 model · Explicit solutions · Weak solutions · Stochastic volatility · Monte Carlo simulations · Option pricing · Non-affine volatility

1 Introduction

Recent work by Kouritzin [17] shows that it is possible to obtain an explicit weak solution for the Heston model, and that this solution can be used to simulate asset prices efficiently. Exploiting the form of the weak solution, which naturally leads to importance sampling, Kouritzin and MacKay [18] suggest the use of sequential sampling algorithms to reduce the variance of the estimator, inspired by the particle

Partial funding in support of this work was provided by FRQNT research support for new academics (number 253180) and by NSERC discovery grants (numbers 04869 and 203089).

I. R. Kouarfate · A. MacKay (✉)
Department of Mathematics, Université du Québec à Montréal, Montreal, QC, Canada
e-mail: kouarfate.iro_rene@courrier.uqam.ca; mackay.anne@uqam.ca

M. A. Kouritzin
Department of Mathematical and Statistical Sciences, University of Alberta, Edmonton, AB, Canada
e-mail: michaelk@ualberta.ca

D. Hernández-Hernández et al. (eds.), *Advances in Probability and Mathematical Statistics*, Progress in Probability 79, https://doi.org/10.1007/978-3-030-85325-9_8

filtering literature. Herein, we show that the main results of [17] can easily be adapted to the 3/2 stochastic volatility model and thus be exploited to develop an efficient simulation algorithm that can be used to price exotic options.

The 3/2 model is a non-affine stochastic volatility model whose analytical tractability was studied in [15] and [19]. A similar process was used in [1] to model stochastic interest rates. Non-affine stochastic volatility models have been shown to provide a good fit to empirical market data, sometimes better than some affine volatility models; see [3] and the references provided in the literature review section of [22]. The 3/2 model in particular is preferred by Carr and Sun [8] as it naturally emerges from consistency requirements in their proposed framework, which models variance swap rates directly.

As a result of the empirical evidence in its favor, and because of its analytical tractability, the 3/2 model has gained traction in the academic literature over the past decade. In particular, Itkin and Carr [16] price volatility swaps and options on swaps for a class of Levy models with stochastic time change and use the 3/2 model as a particular case. The 3/2 model also allows for analytical expressions for the prices of different volatility derivatives; see for example [10, 12, 21]. Chan and Platen [9] consider the 3/2 model for pricing long-dated variance swaps under the real world measure. Zheng and Zeng [22] obtain a closed-form partial transform of a relevant density and use it to price variance swaps and timer options. In [13], the 3/2 model is combined with the Heston model to create the new 4/2 model.

For the 3/2 model's growing popularity, there are very few papers that focus on its simulation. One of them is Baldeaux [4], who adapts the method of [7] to the 3/2 model and suggests variance reduction techniques. The capacity to simulate price and volatility paths from a given market model is necessary in many situations, from pricing exotic derivatives to developing hedging strategies and assessing risk. The relatively small size of the literature concerning the simulation of the 3/2 model could be due to its similarity with the Heston model, which allows for easy transfer of the methods developed for the Heston model to the 3/2 one. Indeed, the 3/2 model is closely linked to the Heston model; the stochastic process governing the variance of the asset price in the 3/2 is the inverse of a square-root process, that is, the inverse of the variance process under Heston.

This link between the Heston and the 3/2 model motivates the present work; Kouritzin [17] mentions that his method cannot survive the spot volatility reaching 0. Since the volatility in the 3/2 model is given by the inverse of a "Heston volatility" (that is, the inverse of a square-root process), it is necessary to restrict the volatility parameters in such a way that the Feller condition is met, in order to keep the spot volatility from exploding. In other words, by definition of the 3/2 model, the variance process always satisfies the Feller condition, which makes it perfectly suitable to the application of the explicit weak solution simulation method of [17].

It is also worthwhile to note that Kouritzin and MacKay [18] notice that the resulting simulation algorithm performs better when the Heston parameters keep the variance process further from 0. It is reasonable to expect that calibrating the 3/2 model to market data give such parameters, since they would keep the variance process (i.e. the inverse of the Heston variance) from reaching very high values. This insight further motivates our work, in which we adapt the method of [17] to the 3/2 model.

As stated above, many simulation methods for the Heston model can readily be applied to the 3/2 model. Most of these methods can be divided into two categories; the first type of simulation schemes relies on discretizing the spot variance and the log-price process. Such methods are typically fast, but the discretization induces a bias which needs to be addressed, see [20] for a good overview. Broadie and Kaya [7] proposed an exact simulation scheme which relies on transition density of the variance process and an inversion of the Fourier transform of the integrated variance. While exact, this method is slow, and has thus prompted several authors to propose approximations and modifications to the original algorithm to speed it up (see for example [2]). Bégin et al. [5] offers a good review of many existing simulation methods for the Heston model.

The simulation scheme proposed by Kouritzin [17] for the Heston model relies on an explicit weak solution for the stochastic differential equation (SDE) describing the Heston model. This result leads to a simulation and option pricing algorithm which is akin to importance sampling. Each path is simulated using an artificial probability measure, called the reference measure, under which exact simulation is possible and fast. The importance sampling price estimator is calculated under the pricing measure by multiplying the appropriate payoff (a function of the simulated asset price and volatility paths) by a likelihood, which weights each payoff proportionally to the likelihood that the associated path generated from the reference measure could have come from the pricing measure. The likelihood used as a weight in the importance sampling estimator is a deterministic function of the simulated variance process, and is thus easy to compute. The resulting pricing algorithm has been shown to be fast and to avoid the problems resulting from discretization of the variance process.

In this paper, we develop a similar method for the 3/2 model by first obtaining a weak explicit solution for the two-dimensional SDE. We use this solution to develop an option price importance estimator, as well as a simulation and option pricing algorithm. Our numerical experiments show that our new algorithm performs at least as well as other popular algorithms from the literature. We find that the parametrization of the model impacts the performance of the algorithm.

The paper is organized as follows. Section 2 contains a detailed presentation of the 3/2 model as well as our main result. Our pricing algorithm is introduced in Sect. 3, in which we also outline existing simulation techniques, which we use in our numerical experiments. The results of these experiments are given in Sect. 4, and Sect. 5 concludes.

2 Setting and Main Results

We consider a probability space $(\Omega, \mathcal{F}, \mathbb{P})$, where $\mathbb{P}$ denotes a pre-determined risk-neutral measure[1] for the 3/2 model. The dynamics of the stock price under this chosen risk-neutral measure are represented by a two-dimensional process $(S, V) = \{(S_t, V_t), t \geq 0\}$ satisfying

$$\begin{cases} dS_t &= r S_t \, dt + \sqrt{V_t} S_t \rho \, dW_t^{(1)} + \sqrt{V_t} S_t \sqrt{1-\rho^2} \, dW_t^{(2)} \\ dV_t &= \kappa \, V_t (\theta - V_t) \, dt + \varepsilon V_t^{3/2} \, dW_t^{(1)}, \end{cases} \tag{2.1}$$

with $S_0 = s_0 > 0$ and $V_0 = v_0 > 0$, and where $W = \{(W_t^{(1)}, W_t^{(2)}), t \geq 0\}$ is a two-dimensional uncorrelated Brownian motion, r, κ, θ and ε are constants satisfying $\kappa > -\frac{\varepsilon^2}{2}$, and $\rho \in [-1, 1]$. The drift parameter r represents the risk-free rate and ρ represents the correlation between the stock price S and its volatility V.

The restriction $\kappa > -\frac{\varepsilon^2}{2}$ imposed on the parameters keeps the variance process from exploding. This property becomes clear when studying the process $U = \{U_t, t \geq 0\}$ defined by $U_t = \frac{1}{V_t}$ for $t \geq 0$. Indeed, it follows from Itô's lemma that

$$\begin{aligned} dU_t &= \kappa\theta \left(\frac{\kappa + \varepsilon^2}{\kappa\theta} - U_t \right) \, dt - \varepsilon \sqrt{U_t} \, dW_t^{(1)} \\ &= \tilde{\kappa}(\tilde{\theta} - U_t) \, dt + \tilde{\varepsilon} \sqrt{U_t} \, dW_t^{(1)} \end{aligned}$$

where $\tilde{\kappa} = \kappa\theta$, $\tilde{\theta} = \frac{\kappa+\varepsilon^2}{\kappa\theta}$ and $\tilde{\varepsilon} = -\varepsilon$. In other words, with the restriction $\kappa > -\frac{\varepsilon^2}{2}$, U is a square-root process satisfying the Feller condition $\tilde{\kappa}\tilde{\theta} > \frac{\tilde{\varepsilon}^2}{2}$, so that $\mathbb{P}(U_t > 0) = 1$ for all $t \geq 0$.

In order to use results obtained for the Heston model and adapt them to the 3/2 model, we express (2.1) in terms of the inverse of the variance process, U, as follows

$$\begin{cases} dS_t &= r S_t \, dt + \sqrt{U_t^{-1}} S_t \rho \, dW_t^{(1)} + \sqrt{U_t^{-1}} S_t \sqrt{1-\rho^2} \, dW_t^{(2)} \\ dU_t &= \tilde{\kappa}(\tilde{\theta} - U_t) \, dt + \tilde{\varepsilon} \sqrt{U_t} \, dW_t^{(1)}, \end{cases} \tag{2.2}$$

with $S_0 = s_0$ and $U_0 = 1/v_0$.

Although U is a square-root process, (2.2) is of course not equivalent to the Heston model. Indeed, in the Heston model, it is the diffusion term of S, rather than its inverse, that follows a square-root process. However, the ideas of Kouritzin [17] can be exploited to obtain an explicit weak solution to (2.2), which will in turn be used to simulate the process.

---

[1] Since our goal in this work is to develop pricing algorithms, we only consider the risk-neutral measure used for pricing purposes.

It is well-known (see for example [14]) that if $n := \frac{4\tilde{\kappa}\tilde{\theta}}{\tilde{\varepsilon}^2}$ is a positive integer, the square-root process U is equal in distribution to the sum of n squared Ornstein-Uhlenbeck processes. Proposition 1 below relies on this result.

Proposition 1 *Suppose that* $n = \frac{4\tilde{\kappa}\tilde{\theta}}{\tilde{\varepsilon}^2} \in \mathbb{N}^+$ *and let* $W^{(2)}, Z^{(1)}, \ldots, Z^{(n)}$ *be independent standard Brownian motions on some probability space* $(\Omega, \mathcal{F}, \mathbb{P})$. *For* $t \geq 0$, *define*

$$S_t = s_0 \exp\left\{\frac{\rho}{\tilde{\varepsilon}} \log\left(\frac{U_t}{U_0}\right) + \left(r + \frac{\rho\tilde{\kappa}}{\tilde{\varepsilon}}\right) t \right.$$
$$\left. - \left(\frac{\rho}{\tilde{\varepsilon}}\left(\tilde{\kappa}\tilde{\theta} - \tilde{\varepsilon}^2/2\right) + \frac{1}{2}\right) \int_0^t U_s^{-1}\,\mathrm{d}s + \sqrt{1-\rho^2} \int_0^t \sqrt{U_s^{-1}}\,\mathrm{d}W_s^{(2)}\right\},$$

$$U_t = \sum_{i=1}^n \left(Y_t^{(i)}\right)^2,$$

where

$$Y_t^{(i)} = \frac{\tilde{\varepsilon}}{2} \int_0^t e^{-\frac{\tilde{\kappa}}{2}(t-u)}\,\mathrm{d}Z_u^{(i)} + e^{-\frac{\tilde{\kappa}}{2}t} Y_0^{(i)}, \qquad \textit{with } Y_0 = \sqrt{U_0/n}$$

and

$$W_t^{(1)} = \sum_{i=1}^n \int_0^t \frac{Y_u^{(i)}}{\sqrt{\sum_{j=1}^n (Y_u^{(j)})^2}}\,\mathrm{d}Z_u^{(i)}.$$

Let $X = (S, U)$, $W = (W^{(1)}, W^{(2)})$ *and let* $\{\mathcal{F}_t\}_{t\geq 0}$ *be the augmented filtration generated by* $(W^{(2)}, Z^{(1)}, \ldots, Z^{(n)})$. *Then,*

- $W^{(1)}$ *is a standard Brownian motion, and*
- (X, W), $(\Omega, \mathcal{F}, \mathbb{P})$, $\{\mathcal{F}_t\}_{t\geq 0}$ *is a weak solution to* (2.2).

Proof We first observe that $Y^{(i)}$, $i \in \{1, \ldots, n\}$, are independent Ornstein-Uhlenbeck processes, and that by Lévy's characterization, $W^{(1)}$ is a Brownian motion. It follows from an application of Itô's lemma that

$$\begin{aligned}
\mathrm{d}U_t &= \sum_{i=1}^n \left(\frac{\tilde{\varepsilon}^2}{4} - \tilde{\kappa}(Y_t^{(i)})^2\right) \mathrm{d}t + \tilde{\varepsilon} Y_t^{(i)}\,\mathrm{d}Z_t^{(i)} \\
&= \left(\frac{n\tilde{\varepsilon}^2}{4} - \tilde{\kappa} \sum_{i=1}^n (Y_t^{(i)})^2\right) \mathrm{d}t + \tilde{\varepsilon} \sum_{i=1}^n Y_t^{(i)}\,\mathrm{d}Z_t^{(i)} \\
&= \left(\frac{n\tilde{\varepsilon}^2}{4} - \tilde{\kappa} U_t\right) \mathrm{d}t + \tilde{\varepsilon}\sqrt{U_t}\,\mathrm{d}W_t^{(1)},
\end{aligned} \tag{2.3}$$

where the last equality is obtained by multiplying and dividing the second term on the right-hand side by $\sqrt{\sum_{j=1}^{n}(Y_t^{(i)})^2}$. Here, since we work under the assumption that $n = \frac{4\tilde{\kappa}\tilde{\theta}}{\tilde{\epsilon}^2}$, (2.3) can be re-written as

$$\mathrm{d}U_t = \tilde{\kappa}\left(\tilde{\theta} - U_t\right)\,\mathrm{d}t + \tilde{\varepsilon}\sqrt{U_t}\,\mathrm{d}W_t^{(1)}.$$

An application of Itô's lemma to S_t completes the proof. □

An alternative, systematic way to verify the functional form for S_t that avoids our Itô-lemma-based guess and verify technique can be found in [17].

It is likely that for a given market calibration of the 3/2 model, $n = \frac{4\tilde{\kappa}\tilde{\theta}}{\tilde{\epsilon}^2}$ is not an integer. For this reason, a more general result is needed to develop a simulation algorithm based on an explicit weak solution.

We generalize the definition of n and let $n = \max\left(\left\lfloor \frac{4\tilde{\kappa}\tilde{\theta}}{\tilde{\varepsilon}^2} + \frac{1}{2}\right\rfloor, 1\right)$. We further define $\tilde{\theta}_n$ by

$$\tilde{\theta}_n = \frac{n\tilde{\varepsilon}^2}{4\tilde{\kappa}}.$$

It follows that $\tilde{\kappa}\tilde{\theta}_n = \frac{n\tilde{\varepsilon}^2}{4}$.

While U above cannot hit 0 under the Feller condition, it can get arbitrarily close, causing U_t^{-1} to blow up. To go beyond the case $\frac{4\tilde{\kappa}\tilde{\theta}}{\tilde{\varepsilon}^2} \in \mathbb{N}$ treated in Proposition 1, we want to change measures, which is facilitated by stopping U from approaching zero.

This change of measure is needed to readjust the distribution of the paths of U simulated using the (wrong) long-term mean parameter $\tilde{\theta}_n$ and Proposition 1. Indeed, Proposition 1 can be used with $\tilde{\theta}_n$ since $\frac{4\tilde{\kappa}\tilde{\theta}_n}{\tilde{\epsilon}^2} = n$ is an integer. Under the new measure, the adjusted paths have the correct distribution, that is, the one associated with the desired parameter $\tilde{\theta}$. This idea is made more precise below.

Given a filtered probability space $(\Omega, \mathcal{F}, \{\mathcal{F}_t\}_{t\geq 0}, \widehat{\mathbb{P}})$ with independent Brownian motions $Z^{(1)}, \ldots, Z^{(n)}$ and $W^{(2)}$, and a fixed $\delta > 0$, we can define $(\widehat{S}, \widehat{U}) = \{(\widehat{S}_t, \widehat{U}_t)\}_{t\geq 0}$ by

$$\widehat{S}_t = s_0 \exp\left\{\frac{\rho}{\tilde{\varepsilon}}\log(\widehat{U}_t/U_0) + \left(r + \frac{\rho\tilde{\kappa}}{\tilde{\varepsilon}}\right)t \right.$$
$$\left. - \left(\frac{\rho}{\tilde{\varepsilon}}\left(\tilde{\kappa}\tilde{\theta} - \tilde{\varepsilon}^2/2\right) + \frac{1}{2}\right)\int_0^t \widehat{U}_s^{-1}\,\mathrm{d}s + \sqrt{1-\rho^2}\int_0^t \widehat{U}_s^{-1/2}\,\mathrm{d}W_s^{(2)}\right\}, \tag{2.4}$$

$$\widehat{U}_t = \sum_{i=1}^{n}(Y_t^{(i)})^2, \tag{2.5}$$

and $\tau_\delta = \inf\{t \geq 0 : \widehat{U}_t \leq \delta\}$, where

$$Y_t^{(i)} = \frac{\tilde{\varepsilon}}{2} \int_0^{t \wedge \tau_\delta} e^{-\frac{\tilde{\kappa}}{2}(t-u)}\, \mathrm{d}Z_u^{(i)} + e^{-\frac{\tilde{\kappa}}{2}(t \wedge \tau_\delta)} Y_0^{(i)}, \qquad \text{with } Y_0 = \sqrt{U_0/n} \tag{2.6}$$

for $i \in \{1, \ldots, n\}$.

Theorem 1, to follow immediately, shows that it is possible to construct a probability measure on $(\Omega, \mathcal{F})$ under which $(\widehat{S}, \widehat{U})$ satisfies (2.2) until $\widehat{U}$ drops below a pre-determined threshold δ.

Theorem 1 *Let $(\Omega, \mathcal{F}, \{\mathcal{F}_t\}_{t \geq 0}, \widehat{\mathbb{P}})$ be a filtered probability space on which $Z^{(1)}, \ldots, Z^{(n)}, W^{(2)}$ are independent Brownian motions. Let $(\widehat{S}, \widehat{U})$ be defined as in (2.4) and (2.5) and let $\tau_\delta = \inf\{t \geq 0 : \widehat{U}_t \leq \delta\}$ for some $\delta \in (0, 1)$. Define*

$$\widehat{L}_t^{(\delta)} = \exp\left\{ -\frac{\tilde{\kappa}(\tilde{\theta}_n - \tilde{\theta})}{\tilde{\varepsilon}} \int_0^t \widehat{U}_v^{-1/2}\, \mathrm{d}\widehat{W}_v^{(1)} - \frac{\tilde{\kappa}^2}{2} \left(\frac{\tilde{\theta}_n - \tilde{\theta}}{\tilde{\varepsilon}} \right)^2 \int_0^t \widehat{U}_v^{-1}\, \mathrm{d}v \right\} \tag{2.7}$$

with

$$\widehat{W}_t^{(1)} = \sum_{i=1}^n \int_0^t \frac{Y_u^{(i)}}{\sqrt{\sum_{j=1}^n (Y_u^{(j)})^2}}\, \mathrm{d}Z_u^{(i)} \tag{2.8}$$

and $\mathbb{P}^\delta(A) = \widehat{E}[1_A \widehat{L}_T^{(\delta)}]\ \forall A \in \mathcal{F}_T$ for $T > 0$.

Then, under the probability measure $\mathbb{P}^\delta$, $(W^{(1)}, W^{(2)})$, where

$$W_t^{(1)} = \widehat{W}_t^{(1)} + \tilde{\kappa} \frac{\tilde{\theta} - \tilde{\theta}_n}{\tilde{\varepsilon}} \int_0^{t \wedge \tau_\delta} \widehat{U}_s^{-1/2}\, \mathrm{d}s,$$

are independent Brownian motions and $(\widehat{S}, \widehat{U})$ satisfies

$$\mathrm{d}\widehat{S}_t = \begin{cases} r\widehat{S}_t\, \mathrm{d}t + \widehat{U}_t^{-1/2} \widehat{S}_t \rho\, \mathrm{d}W_t^{(1)} + \widehat{U}_t^{-1/2} \widehat{S}_t \sqrt{1-\rho^2}\, \mathrm{d}W_t^{(2)}, & t \leq \tau_\delta \\ r_\delta \widehat{S}_t\, \mathrm{d}t + \sigma_\delta \widehat{S}_t\, \mathrm{d}W_t^{(2)}, & t > \tau_\delta, \end{cases}$$

$$\mathrm{d}\widehat{U}_t = \begin{cases} \tilde{\kappa}(\tilde{\theta} - \widehat{U}_t)\, \mathrm{d}t + \tilde{\varepsilon} \widehat{U}_t^{1/2}\, \mathrm{d}W_t^{(1)}, & t \leq \tau_\delta \\ 0, & t > \tau_\delta \end{cases} \tag{2.9}$$

on $[0, T]$, with

$$r_\delta = r + \frac{\rho}{2\tilde{\varepsilon}\delta} \left(2\tilde{\kappa}\delta - 2\tilde{\kappa}\tilde{\theta} + \tilde{\varepsilon}^2 - \rho\tilde{\varepsilon}^2 \right),$$

$$\sigma_\delta = \sqrt{\frac{1-\rho^2}{\delta}}.$$

Proof Let $D = \mathcal{S}(\mathbb{R}^2)$, the rapidly decreasing functions. They separate points and are closed under multiplication so they separate Borel probability measures (see [6]) and hence are a reasonable martingale problem domain.

To show that $(\widehat{X}, W)$, $(\Omega, \mathcal{F}, \mathbb{P}^\delta)$, $\{\widehat{\mathcal{F}}_t\}_{t\geq 0}$, with $\widehat{X} = (\widehat{S}, \widehat{U})$ and $W = (W^{(1)}, W^{(2)})$, is a solution to (2.9), we show that it solves the martingale problem associated with the linear operator

$$\mathcal{A}_t f(s,u) = \left(rsf_s(s,u) + \tilde{\kappa}(\tilde{\theta} - u) f_u(s,u) + \frac{1}{2}s^2u^{-1} f_{ss}(s,u) + \rho\tilde{\varepsilon} s f_{su}(s,u) \right.$$
$$\left. + \frac{1}{2}\varepsilon^2 u f_{uu}(s,u)\right) \mathbb{1}_{[0,\tau_\delta]}(t) + \left(r_\delta s f_s(s,u) + \frac{1-\rho^2}{2\delta^2} f_{ss} s^2\right) \mathbb{1}_{[\tau_\delta, T]}(t)$$

where $f_s = \frac{\partial f(s,u)}{\partial s}$, $f_u = \frac{\partial f(s,u)}{\partial u}$, $f_{ss} = \frac{\partial^2 f(s,u)}{\partial s^2}$, $f_{uu} = \frac{\partial^2 f(s,u)}{\partial u^2}$ and $f_{su} = \frac{\partial^2 f(s,u)}{\partial s \partial u}$. That is, we show that for any function $f \in D$, the process

$$M_t(f) = f(\widehat{S}_t, \widehat{U}_t) - f(\widehat{S}_0, \widehat{U}_0) - \int_0^t (\mathcal{A}_s f)(\widehat{S}_v, \widehat{U}_v)\, \mathrm{d}v,$$

is a continuous, local martingale.

First, we note by (2.4), (2.5), (2.6) as well as Itô's lemma that $(\widehat{S}, \widehat{U})$ satisfies a two-dimensional SDE similar to the 3/2 model (2.2), but with parameters κ, θ_n, r_δ and $\widehat{r}_t = r - \frac{\tilde{\kappa}\rho}{\tilde{\varepsilon}}(\tilde{\theta} - \tilde{\theta}_n)\widehat{U}_t^{-1}$. That is, $((\widehat{S}, \widehat{U}), \widehat{W})$, $(\Omega, \mathcal{F}, \widehat{\mathbb{P}})$, $\{\widehat{\mathcal{F}}_t\}_{t\geq 0}$, where $\{\widehat{\mathcal{F}}_t\}_{t\geq 0}$ is the augmented filtration generated by $(Z_1, \ldots, Z_n, W^{(2)})$, is a solution to

$$\mathrm{d}\widehat{S}_t = \begin{cases} \widehat{r}_t \widehat{S}_t\, \mathrm{d}t + \widehat{U}_t^{-1/2} \widehat{S}_t \rho\, \mathrm{d}\widehat{W}_t^{(1)} + \widehat{U}_t^{-1/2} \widehat{S}_t \sqrt{1-\rho^2}\, \mathrm{d}W_t^{(2)}, & t \leq \tau_\delta, \\ r_\delta \widehat{S}_t\, \mathrm{d}t + \sigma_\delta \widehat{S}_t\, \mathrm{d}W_t^{(2)}, & t > \tau_\delta, \end{cases} \tag{2.10}$$

$$\mathrm{d}\widehat{U}_t = \begin{cases} \tilde{\kappa}(\tilde{\theta}_n - \widehat{U}_t)\, \mathrm{d}t + \tilde{\varepsilon} \widehat{U}_t^{1/2}\, \mathrm{d}\widehat{W}_t^{(1)}, & t \leq \tau_\delta, \\ 0, & t > \tau_\delta. \end{cases} \tag{2.11}$$

with $\widehat{S}_0 = s_0$, $\widehat{U}_0 = 1/v_0$ and $\widehat{W}^{(1)}$ defined by (2.8). It follows that for any function $f \in D$,

$$df(\widehat{S}_t, \widehat{U}_t) = \mathcal{L}_t^{(\delta)} f(\widehat{S}_t, \widehat{U}_t) dt$$
$$+ \left(\rho \widehat{S}_t \widehat{U}_t^{-1/2} f_s(\widehat{S}_t, \widehat{U}_t) + \tilde{\varepsilon} \widehat{U}_t^{1/2} f_u(\widehat{S}_t, \widehat{U}_t)\right) \mathbb{1}_{[0,\tau_\delta]}(t)\, \mathrm{d}\widehat{W}_t^{(1)}$$
$$+ \left(\widehat{U}_t^{-1/2} \mathbb{1}_{[0,\tau_\delta]}(t) + \delta^{-1/2} \mathbb{1}_{[\tau_\delta, T]}\right) \sqrt{1-\rho^2} \widehat{S}_t f_s(\widehat{S}_t, \widehat{U}_t)\, \mathrm{d}W_t^{(2)}, \tag{2.12}$$

where the linear operator $\mathcal{L}^{(\delta)}$ is defined by

$$\mathcal{L}_t^{(\delta)} f(s,u) = \left(\widehat{r}_t s f_s(s,u) + \tilde{\kappa}(\tilde{\theta}_n - u) f_u(s,u) + \frac{1}{2} s^2 u^{-1} f_{ss}(s,u) + \rho\tilde{\varepsilon} s f_{su}(s,u) + \frac{1}{2}\varepsilon^2 u f_{uu}(s,u)\right) \mathbb{1}_{[0,\tau_\delta]}(t) + \left(r_\delta s f_s(s,u) + \frac{1-\rho^2}{2\delta^2} f_{ss} s^2\right) \mathbb{1}_{[\tau_\delta,T]}(t). \tag{2.13}$$

We observe that $\widehat{L}_t^{(\delta)}$ satisfies the Novikov condition, since by definition of $\widehat{U}_t$,

$$\frac{|\tilde{\kappa}(\tilde{\theta}_n - \tilde{\theta})|^2}{\tilde{\varepsilon}^2 \widehat{U}_t} \leq \frac{|\tilde{\kappa}(\tilde{\theta}_n - \tilde{\theta})|^2}{\tilde{\varepsilon}^2 \delta},$$

$\widehat{\mathbb{P}}$-a.s. for all $t \geq 0$. It follows that $\widehat{L}_t^{(\delta)}$ is a martingale and that $\mathbb{P}^\delta$ is a probability measure.

We also have from (2.7) and (2.12) that for $f(s,u) \in C^2([0,\infty]^2)$,

$$\left[\widehat{L}^{(\delta)}, f(\widehat{S},\widehat{U})\right]_t = \int_0^{t\wedge\tau_\delta} \widehat{L}_v^{(\delta)} \left((r-\widehat{r}_v)\widehat{S}_v f_s(\widehat{S}_v,\widehat{U}_v) + \tilde{\kappa}(\tilde{\theta} - \tilde{\theta}_n) f_u(\widehat{S}_v,\widehat{U}_v)\right) \mathrm{d}v. \tag{2.14}$$

Next we define the process $\widehat{M}(f)$ for any $f \in D$ by

$$\begin{aligned}\widehat{M}_t(f) &= \widehat{L}_t^{(\delta)} f(\widehat{S}_t,\widehat{U}_t) - \widehat{L}_0^{(\delta)} f(\widehat{S}_0,\widehat{U}_0) - \int_0^t \widehat{L}_v^{(\delta)} \mathcal{A}_v f(\widehat{S}_v,\widehat{U}_v)\,\mathrm{d}v \\ &= \widehat{L}_t^{(\delta)} f(\widehat{S}_t,\widehat{U}_t) - \widehat{L}_0^{(\delta)} f(\widehat{S}_0,\widehat{U}_0) - \left[\widehat{L}^{(\delta)}, f(\widehat{S},\widehat{U})\right]_t - \int_0^t \widehat{L}_v^{(\delta)} \mathcal{L}_v^{(\delta)} f(\widehat{S}_v,\widehat{U}_v)\,\mathrm{d}v.\end{aligned} \tag{2.15}$$

Using integration by parts, we obtain

$$\begin{aligned}\widehat{M}_t(f) &= \int_0^t \widehat{L}_v^{(\delta)}\,\mathrm{d}f(\widehat{S}_v,\widehat{U}_v)\,\mathrm{d}v + \int_0^t f(\widehat{S}_v,\widehat{U}_v)\,\mathrm{d}\widehat{L}_v^{(\delta)} - \int_0^t L_v^{(\delta)}\mathcal{L}_v^{(\delta)} f(\widehat{S}_v,\widehat{U}_v)\,\mathrm{d}v \\ &= \int_0^t \widehat{L}_v^{(\delta)} \Big[\tilde{\kappa}(\tilde{\theta} - \tilde{\theta}_n)\widehat{U}_v^{-1/2} f(\widehat{S}_v,\widehat{U}_v) + \Big(\rho\widehat{S}_v\widehat{U}_v^{-1/2} f_s(\widehat{S}_v,\widehat{U}_v) \\ &\quad + \tilde{\varepsilon}\widehat{U}_v^{1/2} f_u(\widehat{S}_v,\widehat{U}_v)\Big) \mathbb{1}_{[0,\tau_\delta]}(v)\Big]\,\mathrm{d}\widehat{W}_v^{(1)} \\ &\quad + \int_0^t \widehat{L}_v^{(\delta)} \left(\widehat{U}_v^{-1/2}\mathbb{1}_{[0,\tau_\delta]}(v) + \delta^{-1/2}\mathbb{1}_{[\tau_\delta,T]}(v)\right)\sqrt{1-\rho^2}\,\widehat{S}_v f_s(\widehat{S}_v,\widehat{U}_v)\,\mathrm{d}W_v^{(2)}\end{aligned}$$

so $\widehat{M}_t(f)$ is a local martingale. However, since f is rapidly decreasing, $sf_s(s,u)$, $uf_u(s,u)$, $sf_{su}(s,u)$ and $uf_{uu}(s,u)$ are all bounded. We also have that $\widehat{U}_v \geq \delta$ and

$\widehat{L}_v^{(\delta)}$ is integrable for all v. Hence, it follows by (2.13), (2.14), (2.15) and Tonelli that $\widehat{M}(f)$ is a martingale.

To finish the proof, it suffices to follow the remark on p.174 of [11] and show that

$$E\left[\left(f(\widehat{S}_{t_{n+1}}, \widehat{U}_{t_{n+1}}) - f(\widehat{S}_{t_n}, \widehat{U}_{t_n}) - \int_{t_n}^{t_{n+1}} \mathcal{A}_v f(\widehat{S}_v, \widehat{U}_v)\,\mathrm{d}v\right) \prod_{k=1}^{n} h_k(\widehat{S}_{t_k}, \widehat{U}_{t_k})\right] = 0, \tag{2.16}$$

for $0 \leq t_1 < t_2 < \ldots < t_{n+1}$, $f \in D$, $h \in B(\mathbb{R}^2)$ (the bounded, measurable functions) and where $\widehat{E}[\cdot]$ denotes the $\widehat{\mathbb{P}}$-expectation. To do so, we re-write the left-hand side of (2.16) as

$$\begin{aligned}
&\widehat{E}\left[\widehat{L}_{t_{n+1}}^{(\delta)}\left(f(\widehat{S}_{t_{n+1}}, \widehat{U}_{t_{n+1}}) - f(\widehat{S}_{t_n}, \widehat{U}_{t_n}) - \int_{t_n}^{t_{n+1}} \mathcal{A}_v f(\widehat{S}_v, \widehat{U}_v)\,\mathrm{d}v\right) \prod_{k=1}^{n} h_k(\widehat{S}_{t_k}, \widehat{U}_{t_k})\right] \\
&= \widehat{E}\left[\left(\widehat{L}_{t_{n+1}}^{(\delta)} f(\widehat{S}_{t_{n+1}}, \widehat{U}_{t_{n+1}}) - \widehat{L}_{t_n}^{(\delta)} f(\widehat{S}_{t_n}, \widehat{U}_{t_n}) - \int_{t_n}^{t_{n+1}} \widehat{L}_v^{(\delta)} \mathcal{A}_v f(\widehat{S}_v, \widehat{U}_v)\,\mathrm{d}v\right) \prod_{k=1}^{n} h_k(\widehat{S}_{t_k}, \widehat{U}_{t_k})\right] \\
&= \widehat{E}\left[\left(\widehat{M}_{t_{n+1}}(f) - \widehat{M}_{t_n}(f)\right) \prod_{k=1}^{n} h_k(\widehat{S}_{t_k}, \widehat{U}_{t_k})\right],
\end{aligned}$$

which is equal to 0 since $\widehat{M}(f)$ is a martingale. We can then conclude that $(\widehat{S}, \widehat{U})$ solves the martingale problem for $\mathcal{A}$ with respect to $\widehat{\mathbb{P}}$. □

Remark 1 In Theorem 1, we indicate the dependence of the process $\widehat{L}^{(\delta)}$ on the threshold δ via the superscript. Indeed, $\widehat{L}^{(\delta)}$ depends on δ through $\widehat{U}$. Going forward, for notational convenience, we drop the superscript, keeping in mind the dependence of the likelihood process on δ.

3 Pricing Algorithm

In this section, we show how Theorem 1 can be exploited to price a financial option in the 3/2 model. First, we justify that $(\widehat{S}, \widehat{U})$ defined in (2.4) and (2.5) can be used to price an option in the 3/2 model, even if they satisfy (2.2) only up to τ_δ. We also present an algorithm to simulate paths of $(\widehat{S}, \widehat{U})$ under the 3/2 model as well as the associated importance sampling estimator for the price of the option.

3.1 Importance Sampling Estimator of the Option Price

For the rest of this paper, we consider an option with maturity $T \in \mathbb{R}^+$ whose payoff can depend on the whole path of $\{(S_t, V_t)\}_{t\in[0,T]}$, or equivalently, $\{(S_t, U_t)\}_{t\in[0,T]}$. Indeed, since $V_t = U_t^{-1}$ for all $0 \le t \le T$ and to simplify exposition, we will keep on working in terms of U, the inverse of the variance process, going forward. We consider a *payoff function* $\phi_T(S, U)$ with $E[|\phi_T(S, U)|] < \infty$. We call $\pi_0 = E[\phi_T(S, U)]$ the *price of the option* and the function ϕ_T, its discounted payoff. For example, a call option, which pays out the difference between the stock price at maturity, S_T, and a pre-determined exercise price K if this difference is positive, has discounted payoff function $e^{-rT}\max(S_T - K, 0)$ and price $E[e^{-rT}\max(S_T - K, 0)]$.

Remark 2 We work on a finite time horizon and the option payoff function ϕ_T only depends on (S, U) up to T. We use the index T to indicate this restriction on (S, U).

The next proposition shows that it is possible to use $(\widehat{S}, \widehat{U})$, rather than (S, U), to price an option in the 3/2 model.

Proposition 2 *Suppose (S, U) is a solution to the 3/2 model (2.2) on probability space $(\Omega, \mathcal{F}, \mathbb{P})$ and $\tau_\delta = \inf\{t \ge 0 : U_t \le \delta\}$. Define $(\widehat{S}, \widehat{U})$ by (2.4) and (2.5), set $\widehat{\tau}_\delta = \inf\{t \ge 0 : \widehat{U}_t \le \delta\}$ for $\delta \in (0, 1)$ and let $\phi_T(S, U)$ be a payoff function satisfying $E[|\phi_T(S, U)|] < \infty$. Then,*

$$\lim_{n\to\infty} E^{1/n}[\phi_T(\widehat{S}, \widehat{U})\mathbb{1}_{\hat{\tau}_{1/n} > T\}}] = E[\phi_T(S, U)],$$

where $E^\delta[\cdot]$ denotes the expectation under the measure $\mathbb{P}^\delta$ defined in Theorem 1.

Proof By Theorem 1, $(\widehat{S}, \widehat{U})$ satisfies (2.2) on $[0, \hat{\tau}_{1/n}]$ under the measure $\mathbb{P}^{1/n}$. It follows that

$$E^{1/n}[\phi_T(\widehat{S}, \widehat{U})\mathbb{1}_{\{\widehat{\tau}_{1/n} > T\}}] = E[\phi_T(S, U)\mathbb{1}_{\{\tau_{1/n} > T\}}].$$

Because U satisfies the Feller condition, $\lim_{n\to\infty} \mathbb{1}_{\{\tau_{1/n} \le T\}} = 0$, $\mathbb{P}$-a.s. and

$$\lim_{n\to\infty} E^{1/n}[\phi_T(\widehat{S}, \widehat{U})\mathbb{1}_{\{\widehat{\tau}_{1/n} > T\}}] = \lim_{n\to\infty} E[\phi_T(S, U)\mathbb{1}_{\{\tau_{1/n} > T\}}] = E[\phi_T(S, U)]$$

by the dominated convergence theorem.

□

We interpret Proposition 2 in the following manner: by choosing δ small enough, it is possible to approximate π_0 by $\pi_0^{(\delta)} := E^\delta[\phi_T(\widehat{S}, \widehat{U})\mathbb{1}_{\{\tau_\delta > T\}}]$, that is, using $(\widehat{S}, \widehat{U})$ rather than (S, U). The advantage of estimating the price of an option via $(\widehat{S}, \widehat{U})$ is that the trajectories can easily be simulated exactly under the reference measure $\widehat{\mathbb{P}}$ defined in Theorem 1. In practice, we will show in Sect. 4 that for reasonable 3/2 model calibrations, it is usually possible to find δ small enough that $E^{\delta}[\phi_T(\widehat{S}, \widehat{U})\mathbb{1}_{\{\hat{\tau}_\delta > T\}}]$ is almost undistinguishable from π_0.

In the rest of this section, we explain how $\pi_0^{(\delta)}$ can be approximated with Monte Carlo simulation. As mentioned above, paths of $(\widehat{S}, \widehat{U})$ are easily simulated under the reference measure $\widehat{\mathbb{P}}$, not under $\mathbb{P}^{\delta}$. It is therefore necessary to express $\pi_0^{(\delta)}$ using Theorem 1 in the following manner

$$\pi_0^{(\delta)} = E^{\delta}[\phi_T(\widehat{S}, \widehat{U})\mathbb{1}_{\{\tau_\delta > T\}}] = \widehat{E}[\widehat{L}_T\, \phi_T(\widehat{S}, \widehat{U})\mathbb{1}_{\{\tau_\delta > T\}}]. \tag{3.1}$$

From (3.1) and the strong law of large numbers, we can define $\widehat{\pi}_0^{(\delta)}$, an importance estimator for $\pi_0^{(\delta)}$, by

$$\widehat{\pi}_0^{(\delta)} = \frac{\sum_{j=1}^N \phi_T(\widehat{S}^{(j)}, \widehat{U}^{(j)})\widehat{L}_T^{(j)}\mathbb{1}_{\{\tau_\delta^{(j)} > T\}}}{\sum_{j=1}^N \widehat{L}_T^{(j)}}, \tag{3.2}$$

where $\left\{\widehat{S}^{(j)}, \widehat{U}^{(j)}, \widehat{L}^{(j)}\right\}_{j=1}^N$ are $N \in \mathbb{N}$ simulated paths of $(\widehat{S}, \widehat{U}, \widehat{L})$.

3.2 Simulating Sample Paths

In light of Proposition 2, we now focus on the simulation of $(\widehat{S}_t, \widehat{U}_t, \widehat{L}_t)_{t \le \tau_\delta}$. Using (2.4) and (2.6), $\widehat{S}$ and Y can easily be discretized for simulation purposes. To simplify the simulation of the process $\widehat{L}$, we write (2.7) as a deterministic function of $\widehat{U}$ in Proposition 3 below.

Proposition 3 *Let $\widehat{L}_t$ be defined as in Theorem 1, with $\widehat{U}$ defined by* (2.5). *Then, for $t \le \tau_\delta$, $\widehat{L}_t$ can be written as*

$$\widehat{L}_t = \exp\left\{\frac{-(\tilde{\kappa}\tilde{\theta}_n - \tilde{\kappa}\tilde{\theta})}{\tilde{\varepsilon}^2}\left[\log(\widehat{U}_t/\widehat{U}_0) + \tilde{\kappa}t + \frac{\tilde{\kappa}\tilde{\theta} - 3\tilde{\kappa}\tilde{\theta}_n + \tilde{\varepsilon}^2}{2}\int_0^t \widehat{U}_s^{-1}\,\mathrm{d}s\right]\right\}. \tag{3.3}$$

Proof An application of Itô's lemma to $\log \widehat{U}_t$ for $t \le \hat{\tau}_\delta$ yields

$$\log(\widehat{U}_t/\widehat{U}_0) = (\tilde{\kappa}\tilde{\theta}_n - \tilde{\varepsilon}^2/2)\int_0^t \widehat{U}_s^{-1}\,\mathrm{d}s - \tilde{\kappa}t + \tilde{\varepsilon}\int_0^t \widehat{U}_s^{-1/2}\,\mathrm{d}\widehat{W}_s^{(1)}. \tag{3.4}$$

Isolating $\int_0^t \widehat{U}_s^{-1/2}\,\mathrm{d}\widehat{W}_s^{(1)}$ in (3.4) and replacing the resulting expression in (2.7) gives the result. □

For $t \in [0, T)$ and $h \in (0, T-t)$, for simulation purposes, we can re-write (2.4), (2.6) and (3.3) in a recursive manner as

$$\widehat{S}_{t+h} = \widehat{S}_t \exp\left\{\frac{\rho}{\bar{\varepsilon}} \log(\widehat{U}_{t+h}/\widehat{U}_t) + ah - b\int_t^{t+h} \widehat{U}_s^{-1}\,\mathrm{d}s + \sqrt{1-\rho^2}\int_t^{t+h} \widehat{U}_s^{-1/2}\,\mathrm{d}W_s^{(2)},\right\} \tag{3.5}$$

$$Y_{t+h}^{(i)} = Y_t^{(i)} e^{-\frac{\tilde{\kappa}}{2}h} + \frac{\tilde{\varepsilon}}{2}\int_t^{(t+h)\wedge\tau_\delta} e^{-\frac{\tilde{\kappa}}{2}(t+h-u)}\,\mathrm{d}Z_u, \qquad \text{for } i = 1, \ldots, n, \tag{3.6}$$

and

$$\widehat{L}_{(t+h)\wedge\tau_\delta} = L_t \exp\left\{c\left(\log(\widehat{U}_{(t+h)\wedge\tau_\delta}/\widehat{U}_t) + \tilde{\kappa}(h \vee (\tau_\delta - t)) + d\int_t^{(t+h)\wedge\tau_\delta} \widehat{U}_s^{-1}\,\mathrm{d}s\right)\right\}, \tag{3.7}$$

where

$$a = r + \frac{\rho\tilde{\kappa}}{\bar{\varepsilon}} \qquad b = \frac{\rho}{\bar{\varepsilon}}(\tilde{\kappa}\tilde{\theta} - \tilde{\varepsilon}^2/2) \qquad c = -\frac{\tilde{\kappa}\tilde{\theta}_n - \tilde{\kappa}\tilde{\theta}}{\varepsilon^2} \qquad d = \frac{\tilde{\kappa}\tilde{\theta} - 3\tilde{\kappa}\tilde{\theta}_n + \tilde{\varepsilon}^2}{2}.$$

We now discuss the simulation of $(\widehat{S}_{t+h}, \widehat{U}_{t+h}, \widehat{L}_{t+h})$ given $(\widehat{S}_t, \widehat{U}_t, \widehat{L}_t)$, as well as $\{Y_t^{(i)}\}_{i=1}^n$. Typically, h will be a small time interval, that is, we consider $h \ll T$. It is easy to see from the above that given $Y_t^{(i)}$, $Y_{t+h}^{(i)}$ follows a Normal distribution with mean $Y_t^{(i)} e^{-\frac{\tilde{\kappa}}{2}h}$ and variance $\frac{\tilde{\varepsilon}^2}{4\tilde{\kappa}}(1 - e^{-h\tilde{\kappa}})$. The simulation of $Y_{t+h}^{(i)}$ given $Y_t^{(i)}$ is thus straightforward. $\widehat{U}_{t+h}$ can then be obtained by (2.5) as the sum of the squares of each $Y_{t+h}^{(i)}$, for $i = 1, \ldots, n$.

Given simulated values $\widehat{U}_{t+h}$ and $\widehat{U}_t$, the term $\int_t^{t+h} \widehat{U}_s^{-1}\,\mathrm{d}s$, which appears in both $\widehat{S}_{t+h}$ and $\widehat{L}_{t+h}$, can be approximated using the trapezoidal rule by letting

$$\int_t^{t+h} \widehat{U}_s^{-1}\,\mathrm{d}s \approx \frac{(\widehat{U}_t^{-1} + \widehat{U}_{t+h}^{-1})}{2}h. \tag{3.8}$$

More precise approximations to this integral can be obtained by simulating intermediate values $\widehat{U}_{t+ih}^{-1}$ for $i \in (0, 1)$ and using other quadrature rules. In [17] and [18], Simpson's $\frac{1}{3}$ rule was preferred. In this section, we use a trapezoidal rule only to simplify the exposition of the simulation algorithm.

Given that $\widehat{U}_{t+h} > \delta$ and once an approximation for the deterministic integral $\int_t^{t+h} \widehat{U}_s^{-1}\,\mathrm{d}s$ is calculated, $\widehat{L}_{t+h}$ can be simulated using (3.7). To generate a value for $\widehat{S}_{t+h}$, it suffices to observe that conditionally on $\{\widehat{U}_s\}_{s\in[t,t+h]}$,

$\int_t^{t+h} \widehat{U}_s^{-1/2} \, dW_s^{(2)}$ follows a Normal distribution with mean 0 and variance $\int_t^{t+h} \widehat{U}_s^{-1} \, ds$.

The resulting algorithm produces N paths of $(\widehat{S}, \widehat{U}, \widehat{L})$ and the stopping times τ_δ associated with each path; it is presented in Algorithm 1, in the appendix. These simulated values are then used in (3.2) to obtain an estimate for the price of an option.

4 Numerical Experiment

4.1 Methods and Parameters

In this section, we assess the performance of the pricing algorithm derived from Theorem 1. To do so, we use Monte Carlo simulations to estimate the price of European call options. These Monte Carlo estimates are compared with the exact price of the option, calculated with the analytical expression available for vanilla options in the 3/2 model (see for example [19] and [8]). More precisely, we consider the discounted payoff function $\phi_T(S, U) = e^{-rT} \max(S_T - K, 0)$ for $K > 0$ representing the exercise price of the option and we compute the price estimate according to (3.2).

The precision of the simulation algorithm is assessed using either the mean square error or the relative mean square error, as indicated. We define the mean square error by

$$MSE = E[(\pi_0 - \widehat{\pi}_0^{(\delta)})^2]$$

and the relative mean square error by

$$RelMSE = \frac{E[(\pi_0 - \widehat{\pi}_0^{(\delta)})^2]}{\pi_0},$$

where π_0 is the exact price of the option and $\widehat{\pi}_0^{(\delta)}$ is the estimate calculated with (3.2). The expectations above are approximated by calculating the estimates a large number of times and taking the mean over all runs.

Throughout this section, we consider the five parameter sets presented in Table 1. Parameter set 1 (PS1) was used in [4]. Parameter set 2 (PS2) was obtained by Drimus [10] via the simultaneous fit of the 3/2 model to 3-month and 6-month S&P500 implied volatilities on July 31, 2009. The three other parameter sets are modifications of PS2: PS3 was chosen so that $\frac{4\tilde{\kappa}\tilde{\theta}}{\tilde{\varepsilon}^2} \in \mathbb{N}$, and PS4 and PS5 were selected to have a higher n. Recalling that $n = \max\left(\left\lfloor \frac{4\tilde{\kappa}\tilde{\theta}}{\tilde{\varepsilon}^2} + \frac{1}{2} \right\rfloor, 1\right)$ represents the number of Ornstein-Uhlenbeck processes necessary to simulate the variance

Table 1 Parameter sets

	S_0	V_0	κ	θ	ε	ρ	r	$4\tilde{\kappa}\tilde{\theta}/\tilde{\varepsilon}^2$
PS1	1	1	2	1.5	0.2	−0.5	0.05	204
PS2	100	0.06	22.84	0.218	8.56	−0.99	0.00	5.25
PS3	100	0.06	18.32	0.218	8.56	−0.99	0.00	5.00
PS4	100	0.06	19.76	0.218	3.20	−0.99	0.00	11.72
PS5	100	0.06	20.48	0.218	3.20	−0.99	0.00	12.00

process, we have that $n = 204$ for PS1, $n = 5$ for PS2 and PS3 and $n = 12$ for PS4 and PS5.

Throughout the numerical experiments, the threshold we use is $\delta = 10^{-5}$. For all parameter sets, the simulated process U never crossed below this threshold. Therefore, any δ below 10^{-5} would have yielded the same results.

4.2 Results

In this section, we present the results of our numerical experiments. We first test the sensitivity of our simulation algorithm to n, the number of Ornstein-Uhlenbeck processes to simulate. We then compare the performance of our algorithm to other popular ones in the literature.

4.2.1 Sensitivity to n

We first test the impact of n on the precision of the algorithm. Such an impact was observed in [18] in the context of the Heston model. To verify whether this also holds for the 3/2 model, we consider the first three parameter sets and price at-the-money (that is, $K = S_0$) European call options. For PS1, we follow [4] and compute the price of a call option with maturity $T = 1$. The exact price of this option is 0.4431. PS2 and PS3 are used to obtain the price of at-the-money call options with $T = 0.5$, with respective exact prices 7.3864 and 7.0422. In all three cases, the length of the time step used for simulation is $h = 0.02$.

Here, we assess the precision of the algorithm using the relative MSE in order to compare all three parameter sets, which yield vastly different prices. The relative quadratic error is approximated by computing the price estimators 20 times, for $N \in \{5000, 10{,}000, 50{,}000\}$ simulations. The integral with respect to time (see step (3) of Algorithm 1) is approximated using $M \in \{2, 4\}$ sub-intervals and Simpson's $\frac{1}{3}$ rule.

The results of Table 2 show that the precision of the simulation algorithm seem to be affected by n. Indeed, as a percentage of the exact price, the MSE of the

Table 2 Relative MSE as a percentage of π_0

	PS1		PS2		PS3	
N	$M = 2$	$M = 4$	$M = 2$	$M = 4$	$M = 2$	$M = 4$
5000	0.271	0.316	0.183	0.225	0.239	0.214
10,000	0.203	0.158	0.111	0.112	0.172	0.143
50,000	0.158	0.135	0.085	0.083	0.067	0.070

Table 3 Exact prices π_0 of European call options

K/S_0	$PS2$	$PS3$	$PS4$	$PS5$
0.95	10.364	10.055	11.657	11.724
1	7.386	7.042	8.926	8.999
1.05	4.938	4.586	6.636	6.710

price estimator is higher for PS1 than for the other parameter sets. This observation becomes clearer as N increases.

We recall that for PS3, $\frac{4\bar{\kappa}\bar{\theta}}{\bar{\varepsilon}^2}$ is an integer, while this is not the case for PS2. It follows that for this latter parameter set, the weights $\widehat{L}_T^{(j)}$ are all different, while they are all equal to 1 for PS3. One could expect the estimator using uneven weights to show a worse performance due to the possible great variance of the weights. However, in this case, both estimators show similar a performance; the algorithm does not seem to be affected by the use of uneven weights.

Finally, Table 2 shows that increasing M may not significantly improve the precision of the price estimator. Such an observation is important, since adding subintervals in the calculation of the time-integral slows down the algorithm. Keeping the number of subintervals low reduces computational complexity of our algorithm, making it more attractive.

4.2.2 Comparison to Other Algorithms

In this section, we compare the performance of our new simulation algorithm for the 3/2 model to existing ones. The first benchmark algorithm we consider is based on a Milstein-type discretization of the log-price and variance process. The second one is based on the quadratic exponential scheme proposed by Andersen [2] as a modification to the method of [7], which we adapted to the 3/2 model. These algorithms are outlined in the appendix.

To assess the relative performance of the algorithms, we price in-the-money ($K/S_0 = 0.95$), at-the-money ($K/S_0 = 1$) and out-of-the-money ($K/S_0 = 1.05$) call options with $T = 1$ year to maturity. The exact prices of the options, which are used to calculate the MSE of the price estimates, are given in Table 3. We consider all parameter sets with the exception of PS1, since this parametrization requires the simulation of 204 Ornstein-Uhlenbeck process, which makes our algorithm excessively slow. Run times for the calculation of the Monte Carlo estimators using $N = 50{,}000$ simulations and $M = 2$ subintervals are reported in Table 4.

Table 4 Run times (in seconds), $N = 5 \times 10^5$, $M = 2$

Parameters	Milstein	Quadratic exponential	Weighted
PS2	0.445	2.222	0.712
PS3	0.446	2.224	0.615
PS4	0.444	2.226	1.314
PS5	0.444	2.222	1.312

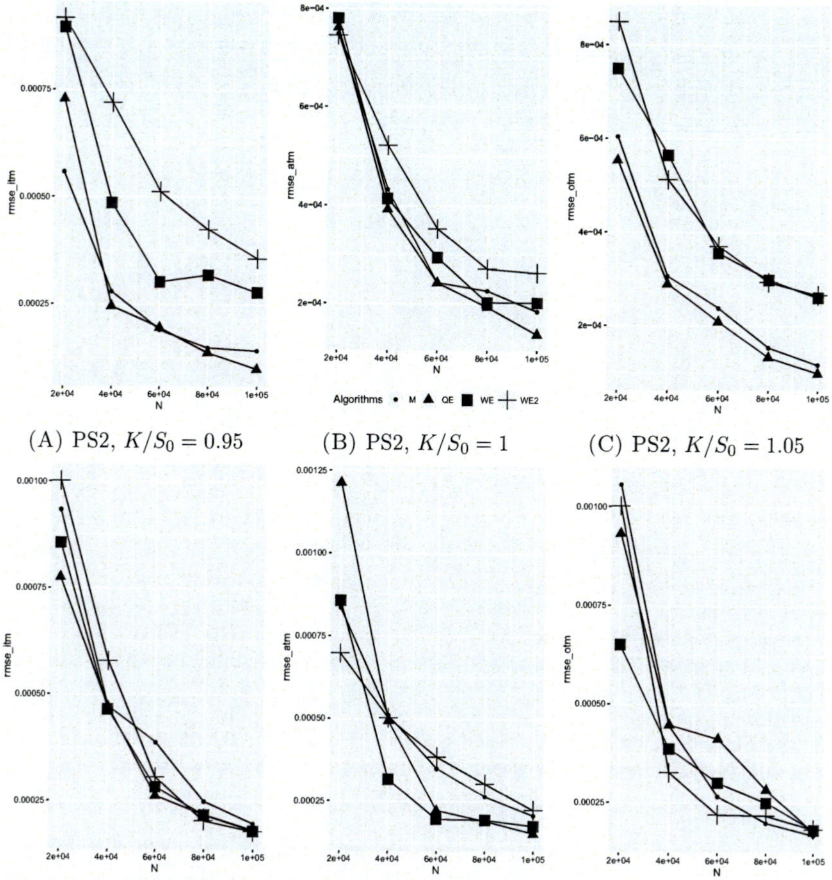

(A) PS2, $K/S_0 = 0.95$ (B) PS2, $K/S_0 = 1$ (C) PS2, $K/S_0 = 1.05$

(D) PS4, $K/S_0 = 0.95$ (E) PS2, $K/S_0 = 1$ (F) PS4, $K/S_0 = 1.05$

Fig. 1 Relative MSE as a function of N, PS2 and PS4, algorithms: Milstein (dot), quadratic exponential (triangle), weighted, $M = 2$ (square), weighted, $M = 4$ (cross)

Figures 1 and 2 present the relative MSE of the price estimator as a function of the number of simulations. We note that the parametrizations considered in Fig. 1 are such that $\frac{4\tilde{\kappa}\tilde{\theta}}{\tilde{\varepsilon}^2} \notin \mathbb{N}$, while the opposite is true for Fig. 2.

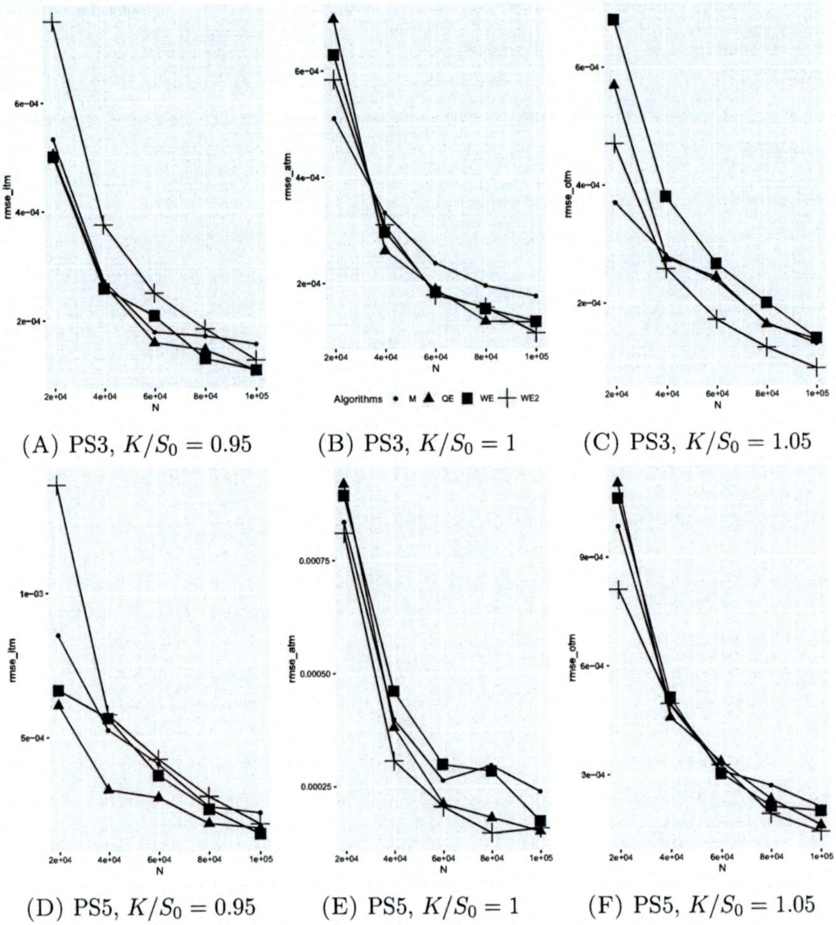

Fig. 2 Relative MSE as a function of N, PS3 and PS5, algorithms: Milstein (dot), quadratic exponential (triangle), weighted, $M = 2$ (square), weighted, $M = 4$ (cross)

Overall, the precision of our weighted simulation algorithm is similar to that of the other two algorithms studied. However, certain parameter sets result in more precise estimates. Figure 1 shows that the MSE is consistently larger with the weighted simulation algorithms than with the benchmark ones for PS2. However, with PS4, the weighted algorithm performs as well as the other two algorithms, or better. We note that for PS2, $n = 5$ while for PS4, $n = 12$. It was observed in [18] in the case of the Heston model that as n increases, the weighted simulation algorithm seems to perform better relatively to other algorithms. This observation also seems to hold in the case of the 3/2 model.

For parametrizations that satisfy $\frac{4\tilde{\kappa}\tilde{\theta}}{\tilde{\varepsilon}^2} \in \mathbb{N}$, such as in Fig. 2, we observe that the weighted simulation algorithm is at least as precise, and often more, than the other algorithms. In this case, all the weights $\widehat{L}_T$ are even, which tends to decrease the variance of the price estimator and thus, to decrease the relative MSE. It is also interesting to note that in the case of Fig. 2, since $\frac{4\tilde{\kappa}\tilde{\theta}}{\tilde{\varepsilon}^2} \in \mathbb{N}$, it is not necessary to simulate $\hat{\tau}_\delta$ and the trajectories $\widehat{L}_T$. Indeed, in this case, it is possible to simplify the algorithm using Proposition 1, which tends to speed it up.

The run times presented in Table 4 show that in general, our method is slower than Milstein's, but faster than the quadratic exponential approximation of [2]. While the run times of the two benchmarks we consider are somewhat constant across the different parametrizations we tested, the speed of our method depends on a two factors; the number of Ornstein-Uhlenbeck processes n to simulate and whether or not $\frac{4\tilde{\kappa}\tilde{\theta}}{\tilde{\varepsilon}^2} \in \mathbb{N}$. This second factor explains the minor differences between the run times reported for PS2 and PS3. However, it should be noted that simulating the weights $\widehat{L}$ is not particularly time consuming, as they are obtained as a deterministic function of $\widehat{U}$ and therefore require no additional simulation. The most significant difference in run times is due to n; for example, it takes twice as long to obtain a price estimate using PS5 ($n = 12, \frac{4\tilde{\kappa}\tilde{\theta}}{\tilde{\varepsilon}^2} \notin \mathbb{N}$) than PS3 ($n = 5, \frac{4\tilde{\kappa}\tilde{\theta}}{\tilde{\varepsilon}^2} \in \mathbb{N}$). While our method is always faster than the one of [2] for the parametrizations studied, we expect that in certain cases (when n is high), it could become slower. Nonetheless, in those cases, our method should be very precise.

We also remark that, when it is used to simulate Heston prices and volatilities, Milstein's method can lead to poor accuracy, especially when the Feller condition is not respected. In the 3/2 model, the Feller condition is always met, so it is normal to expect Milstein's algorithm to perform well. Indeed, Figs. 1 and 2 show that it reaches a similar level of precision as the other methods considered.

5 Conclusion

In this paper, we present a weak explicit solution to the 3/2 model, up until the inverse of the variance process drops below a given threshold. We develop a simulation algorithm based on this solution and show that it can be used to price options in the 3/2 model, since in practice, the inverse variance process stays away from 0. Numerical examples show that our simulation algorithm performs at least as well as popular algorithms presented in the literature. Precision is improved when the parameters satisfy $\frac{4\tilde{\kappa}\tilde{\theta}}{\tilde{\varepsilon}^2} \in \mathbb{N}$ and when n is larger. We also show that it is significantly faster than the quadratic exponential approximation of [2] to the method of [7], which is generally considered to present a good balance between precision and computation time.

It is important to note that the method that we present in this paper could be significantly sped up by the use of sequential resampling, as implemented in [18] for the Heston model. Such improvements, left for future work, could give a significant advantage to our weighted simulation algorithm for the 3/2 model.

Appendix

This section presents the simulation algorithms used to produce the numerical examples in Sect. 4. Algorithm 1 stems from the results we present Theorem 1. Algorithm 2 is a Milstein-type algorithm applied to the 3/2 model. Algorithm 3 is [2]'s approximation to the algorithm proposed by Broadie and Kaya [7], modified for the 3/2 model, since the original algorithm was developed for the Heston model. Algorithms 2 and 3 are considered for comparison purposes.

For all algorithms, we consider a partition $\{0, h, 2h, \ldots, mh\}$, with $mh = T$ of the time interval $[0, T]$, and outline the simulation of N paths of $(\widehat{S}, \widehat{U}, \widehat{L})$, as well as the associated stopping times $\hat{\tau}_\delta$.

To simplify the exposition of Algorithm 1, we define the following constants:

$$\alpha_h = e^{-\frac{\tilde{\kappa}}{2}h}, \qquad \sigma_h = \frac{\tilde{\varepsilon}^2}{4\tilde{\kappa}}\left(1 - e^{-h\tilde{\kappa}}\right).$$

We also drop the hats to simplify the notation.

Algorithm 1 (Weighted explicit simulation)

I. Initialize:

Set the starting values for each simulated path:

$$\{(S_0^{(j)}, L_0^{(j)}, \tau_\delta^{(j)}) = (S_0, 1, T + h)\}_{j=1}^N, \ \{Y_0^{(l,j)} = \sqrt{U_0/n}\}_{l,j=1}^{n,N}$$

II. Loop on time: *for* $i = 1, \ldots, m$

Loop on particles: *for* $j = 1, \ldots, N$, *do*

(1) For $l = 1, \ldots, n$, *generate* $Y_{ih}^{(l,j)}$ *using* $Y_{ih}^{(l,j)} \sim N\left(\alpha_h Y_{(i-1)h}^{(l,j)}, \sigma_h^2\right)$.

(2) Set $U_{ih}^{(j)} = \sum_{l=1}^n (Y_{ih}^{(l,j)})^2$.

(3) Let $IntU_{ih}^{(j)} \approx \int_{(i-1)h}^{ih} (U_s^{(j)})^{-1}\, ds$ *using* (3.8) *(or another quadrature rule).*

(4) Generate $S_{ih}^{(j)}$ *from* $S_{(i-1)h}^{(j)}$ *using* (2.4), *with* $\int_{ih}^{(i-1)h} (U_s^{(j)})^{-1/2} dW_s^{(2)} \sim N(0, IntU_{ih}^{(j)})$.

(5) If $ih \leq \tau_\delta^{(j)}$,

(i) If $U_{ih}^{(j)} > \delta$, *generate* $L_{ih}^{(j)}$ *from* $L_{(i-1)h}^{(j)}$ *using* (3.7).

(ii) Otherwise, set $\tau_\delta^{(j)} = t$.

Algorithm 2 (Milstein)
I. Initialize:

Set the starting values for each simulated path:

$$\{(S_0^{(j)}, U_0^{(j)}) = (S_0, U_0\}_{j=1}^N$$

II. Loop on time: *for* $i = 1, \ldots, m$

Loop on particles: *for* $j = 1, \ldots, N$, *do*

(1) Correct for possible negative values: $\bar{u}^{(j)} = \max(U((i-1)h), 0)$
(2) Generate $U_{ih}^{(j)}$ *from* $U_{(i-1)h}^{(j)}$*:*

$$U_{ih}^{(j)} = U_{(i-1)h}^{(j)} + \tilde{\kappa}(\tilde{\theta} - \bar{u}^{(j)})h + \tilde{\epsilon}\sqrt{\bar{u}^{(j)}h}Z_1^{(j)} + \frac{1}{4}\tilde{\varepsilon}^2((Z_1^{(j)})^2 - 1)h,$$

with $Z_1^{(j)} \sim N(0, 1)$.
(3) Generate $S_{ih}^{(j)}$ *from* $S_{(i-1)h}^{(j)}$*:*

$$S_{ih} = S_{(i-1)h} \exp\left\{\left(r - \frac{1}{2\bar{u}}\right)h + \sqrt{\frac{h}{\bar{u}}}Z_2^{(j)}\right\},$$

with $Z_2^{(j)} \sim N(0, 1)$.

Algorithm 3 (Quadratic exponential)
I. Initialize:

(1) Set the starting values for each simulated path:

$$\{(S_0^{(j)}, U_0^{(j)}) = (S_0, U_0)\}_{j=1}^N$$

(2) Fix the constant $\phi_c \in [1, 2]$.

II. Loop on time: *for* $i = 1, \ldots, m$

Loop on particles: *for* $j = 1, \ldots, N$, *do*

(1) Set the variables $m_{i,j}$ *and* $s_{i,j}$*:*

$$m_{i,j} = \tilde{\theta} + (U_{(i-1)h}^{(j)} - \tilde{\theta})e^{-\tilde{\kappa}h}$$

$$s_{i,j} = \frac{U_{(i-1)h}\tilde{\varepsilon}^2 e^{-\tilde{\kappa}h}}{\tilde{\kappa}}(1 - e^{-\tilde{\kappa}h}) + \frac{\tilde{\theta}\tilde{\varepsilon}^2}{2\tilde{\kappa}}(1 - e^{-\tilde{\kappa}h})^2$$

(2) *Set* $\phi_{i,j} = \frac{s_{i,j}^2}{m_{i,j}^2}$.

(3) *If* $\phi_{i,j} < \phi_c$,
Generate $U_{ih}^{(j)}$ *from* $U_{(i-1)h}^{(j)}$:

$$U_{ih}^{(j)} = a_{i,j}(b_{i,j} + Z^{(j)})^2,$$

where $Z^{(j)} \sim N(0, 1)$ *and*

$$b_{i,j}^2 = 2\phi_{i,j}^{-1} - 1 + \sqrt{2\phi_{i,j}^{-1}}\sqrt{2\phi_{i,j}^{-1} - 1}$$

$$a_{i,j} = \frac{m_{i,j}}{1 + b_{i,j}^2}$$

(4) *If* $\phi_{i,j} \geq \phi_c$,
Generate $U_{ih}^{(j)}$ *from* $U_{(i-1)h}^{(j)}$:

$$U_{ih}^{(j)} = \frac{1}{\beta} \log\left(\frac{1 - p_{i,j}}{1 - X^{(j)}}\right),$$

where $X^{(j)} \sim Uniform(0, 1)$ *and*

$$p_{i,j} = \frac{\psi_{i,j} - 1}{\psi_{i,j} + 1}, \qquad \beta_{i,j} = \frac{1 - p_{i,j}}{m_{i,j}}$$

(5) *Let* $IntU_{ih}^{(j)} \approx \int_{(i-1)h}^{ih} (U_s^{(j)})^{-1}\, ds$ *using* (3.8).

(6) *Generate* $S_{ih}^{(j)}$ *from* $S_{(i-1)h}^{(j)}$ *using* (2.4), *with* $\int_{ih}^{(i-1)h} (U_s^{(j)})^{-1/2} dW_s^{(2)} \sim N(0, IntU_{ih}^{(j)})$.

References

1. Ahn, D.-H., Gao, B.: A parametric nonlinear model of term structure dynamics. Rev. Financ. Stud. **12**(4), 721–762 (1999)
2. Andersen, L.B.: Efficient simulation of the Heston stochastic volatility model. J. Comput. Finance **11**(3), 1–42 (2007)
3. Bakshi, G., Ju, N., Ou-Yang, H.: Estimation of continuous-time models with an application to equity volatility dynamics. J. Financ. Econ. **82**(1), 227–249 (2006)
4. Baldeaux, J.: Exact simulation of the 3/2 model. Int. J. Theor. Appl. Finance **15**(05), 1250032 (2012)
5. Bégin, J.-F., Bédard, M., Gaillardetz, P.: Simulating from the Heston model: A gamma approximation scheme. Monte Carlo Methods Appl. **21**(3), 205–231 (2015)
6. Blount, D., Kouritzin, M.A.: On convergence determining and separating classes of functions. Stoch. Process. Appl. **120**(10), 1898–1907 (2010)

7. Broadie, M., Kaya, O.: Exact simulation of stochastic volatility and other affine jump diffusion processes. Oper. Res. **54**(2), 217–231 (2006)
8. Carr, P., Sun, J.: A new approach for option pricing under stochastic volatility. Rev. Deriv. Res. **10**(2), 87–150 (2007)
9. Chan, L., Platen, E.: Pricing and hedging of long dated variance swaps under a 3/2 volatility model. J. Comput. Appl. Math. **278**, 181–196 (2015)
10. Drimus, G.G.: Options on realized variance by transform methods: a non-affine stochastic volatility model. Quant. Finance **12**(11), 1679–1694 (2012)
11. Ethier, S., Kurtz, T.G.: Markov Processes: Characterization and Convergence. John Wiley & Sons **282** (2009)
12. Goard, J., Mazur, M.: Stochastic volatility models and the pricing of VIX options. Math. Finance Int. J. Math. Stat. Financ. Econ. **23**(3), 439–458 (2013)
13. Grasselli, M.: The 4/2 stochastic volatility model: a unified approach for the Heston and the 3/2 model. Math. Finance **27**(4), 1013–1034 (2017)
14. Hanson, F.B.: Stochastic calculus of Heston's stochastic-volatility model. In: Proceedings of the 19th International Symposium on Mathematical Theory of Networks and Systems–MTNS, vol. 5 (2010)
15. Heston, S.L.: A simple new formula for options with stochastic volatility working paper, Washington University (1997)
16. Itkin, A., Carr, P.: Pricing swaps and options on quadratic variation under stochastic time change models—discrete observations case. Rev. Deriv. Res. **13**(2), 141–176 (2010)
17. Kouritzin, M.A.: Explicit Heston solutions and stochastic approximation for path-dependent option pricing. Int. J. Theor. Appl. Finance **21**(01), 1850006 (2018)
18. Kouritzin, M.A., MacKay, A.: Branching particle pricers with Heston examples. Int. J. Theor. Appl. Finance **23**(01), 2050003 (2020)
19. Lewis, A.: Option valuation under stochastic volatility with mathematica code, Finance Press (2000)
20. Lord, R., Koekkoek, R., Dijk, D.V.: A comparison of biased simulation schemes for stochastic volatility models. Quant. Finan. **10**(2), 177–194 (2010)
21. Yuen, C.H., Zheng, W., Kwok, Y.K.: Pricing exotic discrete variance swaps under the 3/2-stochastic volatility models. Appl. Math. Finance **22**(5), 421–449 (2015)
22. Zheng, W., Zeng, P.: Pricing timer options and variance derivatives with closed-form partial transform under the 3/2 model. Appl. Math. Finance **23**(5), 344–373 (2016)